# Spaces for Highly Mobile People

This book explores how emerging mobility practices have transformed spaces in order to fit the needs of highly mobile people, as well as the changing relationship between people and territory.

It establishes an interdisciplinary and a multiscalar approach to mobility analysis and mobility design through the application of a mobile method of research. Drawing on mobile people in Italy, the book highlights how influential movers appropriate and configure space for their own needs, centring their activities on continuous but distant places and configuring territories with uncertain and evolving limits. This change of perspective allows us to redefine the concept of mobility space, including all the spaces that support the development of emerging mobility practices. It also encourages new perspectives on the way in which the relationship between the individual and territory is evolving into a less sedentary way of inhabiting space.

This book will be of interest to architects, urban scientists and sociologists, as well as postgraduate students who are interested in understanding how mobilities are transforming contemporary cities and territories.

**Bruna Vendemmia** is an architect and urban designer, with a PhD in urbanism. She is post-doctoral fellowship at DAStU Department of Excellence (Politecnico di Milano). Her main research interest focuses on spatial transformations engendered by contemporary changes in mobility practices. In her research she combines qualitative methods with interpretative mapping and spatial analysis to explore the consequences of mobile lives on the configurations of spaces and territorial relations.

## Transport and Mobility

**Series Editor: John Nelson**

The inception of this series marks a major resurgence of geographical research into transport and mobility. Reflecting the dynamic relationships between socio-spatial behaviour and change, it acts as a forum for cutting-edge research into transport and mobility, and for innovative and decisive debates on the formulation and repercussions of transport policy making.

**Mobility Patterns and Urban Structure**
*Paulo Pinho and Cecília Silva*

**Intermodal Freight Terminals**
A Life Cycle Governance Framework
*Jason Monios and Rickard Bergqvist*

**Railway Deregulation in Sweden**
Dismantling a Monopoly
*Gunnar Alexandersson and Staffan Hulten*

**Community-Owned Transport**
*Leigh Glover*

**Geographies of Transport and Mobility**
Prospects and Challenges in an Age of Climate Change
*Stewart Barr, Jan Prillwitz, Tim Ryley and Gareth Shaw*

**Urban Mobilities in the Global South**
*Edited by Tanu Priya Uteng and Karen Lucas*

**Spaces of Congestion and Traffic**
Politics and Technologies in Twentieth-Century London
*David Rooney*

**Gendering Smart Mobilities**
*Edited by Tanu Priya Uteng, Hilda Rømer Christensen, and Lena Levin*

**Spaces for Highly Mobile People**
Emerging Practices of Mobility in Italy
*Bruna Vendemmia*

For more information about this series, please visit: www.routledge.com/Transport-and-Mobility/book-series/ASHSER-1188

# Spaces for Highly Mobile People

## Emerging Practices of Mobility in Italy

**Bruna Vendemmia**

Routledge
Taylor & Francis Group
LONDON AND NEW YORK

First published 2021
by Routledge
2 Park Square, Milton Park, Abingdon, Oxon OX14 4RN

and by Routledge
52 Vanderbilt Avenue, New York, NY 10017

*Routledge is an imprint of the Taylor & Francis Group, an informa business*



*British Library Cataloguing-in-Publication Data*
A catalogue record for this book is available from the British Library

*Library of Congress Cataloging-in-Publication Data*
A catalog record has been requested for this book

ISBN: 978-0-367-40640-0 (hbk)
ISBN: 978-0-367-80821-1 (ebk)

Typeset in Times New Roman
by Taylor & Francis Books

**To Cosma and Bruno**

# Contents

# Figures

# Tables

# Acknowledgments

This research has been founded by the PhD course in Design and Management of Territory XXVII cycle at Politecnico of Milan; the book has been written during a post-doctoral research at DAStU – Excellence Department Politecnico of Milan.

# Foreword

*Paola Pucci*

## Long distance commuters: moving beyond displacement

The book by Bruna Vendemmia focuses on a mobility practice – long distance commuting (LDC) – which has been at the core of a small body of research in recent years.[1]

In current research on long distance commuting, we find a set of shared reasons that originate this mobility, namely fast travel modes, efficient intermodal connections, transformations in the labour market and work programmes, and the spreading of urban populations in peri-urban areas (Ralph 2014; Viry and Kaufmann 2015; Bissell et al. 2017). At the same time, these studies do not offer converging interpretations on the impacts of these mobilities on daily life, the motivations of people involved in this daily travel nor on the attributes and criteria necessary to define long distance commuters.

This makes the book by Bruna Vendemmia a further useful contribution to deepen the knowledge about these work-related mobilities in terms of the conditions that determine them, the consequences in lifestyle and the uses of a territory and its networks.

The interest in focusing on these forms of mobility – despite the fact that these daily mobility practices remain still quantitatively insignificant compared to daily commuting flows – is twofold.

First, by analysing LDC we can read the way in which changes in working patterns increase the complexity of mobility patterns in terms of frequency, time devoted to displacements and distances covered, as well as the combined effects of socio-economical and lifestyles changes. This is due to the fact that practices related to LDC describe more than displacements and, due to their increasingly spatial and temporal variability, they are no longer comparable to traditional categories of commuting. LDCs use space and services in new ways, according to specific times, preferences and constraints that are linked to long daily travel times.

In LDC, mobility is an 'epiphenomenon' of wider transformations in the job market, and transport/communication supply, thus capable of describing social, spatial and temporal changes. Following the interpretation of several scholars, this mobility clearly represents an 'analyser' (Bourdin, 2005), useful for describing transformation in urban life and work-programmes of different profiles of mobile people.

Second, such mobility practices pose interpretative and operational challenges. They question two key principles of the traditional 'utilitarian approach' in transport planning, such as the concept of travel as a derived demand and the understanding of users' objective as travel cost minimisation. This is because, in these mobilities, the increase in travel time budgets is no longer experienced as wasted time, but increasingly as a social and productive time, used to perform different activities during the travel (Lyons and Urry 2005; Jain and Lyons 2008). The same travel choices are guided by much more complex criteria than time cost minimisation: they involve a plurality of aspects linked to both individual (preferences and norms, social ties, household composition) and external factors (transport supply, housing market, labour market).

Following these two perspectives, some considerations may introduce the original and specific contribution offered by this book.

## Exploring the determinants of long distance commuting

Several authors, who pay attention to long distance commuting, have offered various interpretations about the key elements in explaining reasons, dispositions, skills and the lifestyle conditions of LDCs.

In one of the first studies carried out, Öhman and Lindgrend (2003), exploring the determinants of long distance commuting as compared to shorter commuting distances, suggest that mobility strategies and the lifestyle of long distance commuters are characterised by a diversity of driving forces affecting individual and household preferences, along with structural factors. Personal characteristics, preferences and norms, household composition, social ties, labour and housing market conditions and transportation are considered the most influential factors affecting long distance commuter travels (Öhman and Lindgrend 2003:11).

Dealing with the structural factors at the origin of these displacements, Schneider and Meil (2008), as well as Viry and Kaufmann (2015), ascribe LDCs as part of other emerging work-related long distance mobilities, resulting from the combined effects of labour market evolution and a familiarity with the territory and its transport and communication networks that lengthen travel.

For these authors the concept of spatial reversibility explains the LDC experience. This means that LDCs consider their extensive travel as reversible in space and in time, because they use travel time 'to stay in touch with friends and family, and travel speed to be physically present with friends and relatives as much as possible'. In doing so, it is about an 'existential and relational reversibility, compensating for absence by maintaining distant relationship, limiting the impact of absence and limiting the contact with unfamiliar places and unknown people by developing routines' (Viry and Kaufmann, 2015:10).

Reconstructing the societal shifts that may have propelled long distance commuting, Ralph (2014) examined a particular subset of the so-called Eurostar population: the 'Euro-commuters', distinguishing among them three typologies[2] strongly inflected by gender, age, family formation stage, education and specialised training. Becoming an alternative to more conventional forms of migration, this kind of transnational circular mobility depends on an individual's mobility capital and is affected by differing primary motivations, while representing nowadays a prospect for ordinary European professionals to consider as part of their attempts to achieve some sort of durable work-life balance.

In exploring the social life of long distance commuters or 'super-commuters', Bissell, Vannini and Jensen (2016) define them as a 'routine forms of mobility that go beyond the local and urban level, but that are not quite as broad-ranging as international travel' (p. 1). Focusing on the intensity of the lives and mobility experiences of LDC, the authors explore how different intensities of light-touch and forceful events and encounters can create 'a number of bodily capacities accumulated by the LDC to affect and be affected by knowledge, technologies, embodied sensations, and the travelled environment and infrastructures' (Bissell, Vannini and Jensen, 2016:15).

Beyond the thematic and interpretive foci, these studies experiment with a sort of 'ecological approach' to exploring long distance commuting life, dealing with preferences, resources, constrains, social ties and individual goals of people experiencing daily long distance travels.

However, these studies have focused more on 'descriptions of' these phenomena through the use of consolidated research methods. Their resulting findings have not become 'descriptions for better addressing emerging mobility issues' (Pucci and Vecchio, 2019).

## Addressing mobility policy

If research carried out on the LDCs offers important interpretation about the different features of this phenomenon, nevertheless, until now, it has had mainly episodic operational implications for planning and policy.

Approaches focusing on LDC are characterised by in-depth ethnographic and experimental analysis of limited samples, for describing the space-time variability of this mobility, as well as the reasons, behaviours and constraints related to long distance commuting.

Based on this, two conceptual and interpretive limitations affect its potential contribution to planning and policy: the heterogeneous analytical frameworks and their difficult replicability that prevent them from consolidating as ordinary approaches in planning processes and the tension between individual needs and universalistic approaches to them.

If the process of identification of these mobility practices has been tested and updated thanks to qualitative and quantitative approaches, the impacts that they have on the relationship between people and territory is still uncertain, as well as the ability to intercept needs and goods generated by these practices, which influence many different domains of lives and create complex 'commuting spaces' (Bissell, 2018).

Since long distance commuting implies more articulated forms and temporalities than daily commuting, the analysis of LDCs suggests not just the need to combine quantitative and qualitative methodologies, but to propose a way to selectively use the latter in relation to specific conditions, in order to provide an effective framework to address policy measures in terms of service provision and facilities available on the move (services offered in stations and inside vehicles, the organisation and equipment of new spaces of mobility or interventions beyond infrastructure provisions), as well as facilitating the conciliation of individual and family needs (Pucci and Vecchio, 2019).

By addressing the analytical approach in a useful way to explore the dimensions characterising this mobility as a movement in space (displacement), in which the behavioural components are both relevant to address interpretations of this phenomenon and useful for improving the effectiveness of mobility policies, it could be possible to detect specific needs related to this mobility practice, defining which needs require priority for public interventions, recognising the relevance of rising mobility alternatives, but also promoting innovations in the established normative and planning frameworks.

The challenges posed by these mobility practices in terms of interpretative framework, relevant analytical approaches and policy measures have been faced by this book, which deals with the main determinants of a sample of LDCs, analysing the reasons for commuting long distances, the differentiated relationships with LDCs' home spaces, the activities carried out during the travel and the uses of the spaces of movement, proposing some relevant tools for representing and mapping the complex geographies of these practices.

Dealing with the emerging needs, times and conditions of using spaces and networks, as well as the intensity of interactions activated by these practices, thanks to an in-depth fieldwork with a sample of Italian long distance commuters, the book by Vendemmia offers more than a descriptive interpretation of this phenomenon. It assumes a design perspective able to make the travel time of LDCs an 'active' moment, suggesting solutions to provide spaces, equipment and services useful for these practices.

## Notes

1 Among them Öhman and Lindgrend (2003); Kaufmann (2005) ; Schneider and Meil (2008); Sandow and Westin (2010); Ralph, (2014); Viry and Kaufmann (2015); Bissell et al. (2017); Pucci (2019).

2 Among them: 'survivors who undertake this mobility primarily for livelihood purposes; thrivers for the sake of optimizing lifestyles; strivers do so in the service of upward career mobility' (Ralph, 2014:34).

# Introduction

In recent years the availability of ICT and the increasing use of smartphones, tablets and faster means of transport have tended to nullify the impact of spatial distance on travellers. The transformation from door-to-door to place-to-place and finally person-to-person communities (Elliot and Urry, 2010: 48), the changes in the configuration of the nuclear family (Widmer, 2006), the development of networks of distant personal relations (Larsen, Axhausen and Urry, 2006), the massive access of women to work and the diffusion of new forms of job contracts (Colleoni and Zajczyk, 2003: 40), are all factors that have contributed to a radical transformation of mobility practices. Lastly, the recent economic crisis has led, on one hand, to an expansion of job search basins, influencing the 'inclination to mobility' (Ravalet, Vincent-Geslin and Dubois, 2014), and, on the other hand, to the configuration of new job contracts such as 'involuntary part-time'.

The main aim of this book is to study those transformations of mobility practices and to understand how far they have changed the way people use and transform spaces in order to better fit their needs and requirements. The spatial aspects of the increase of mobility are still relatively unexplored, though they are widespread and their analyses become important in improving people's lives. After showing the main goal of the book, in the introduction a short overview of the chapters will be outlined.

## Aim of the research

Nowadays, more and more people are living their lives across a stretched space consisting of work, family and friendship networks, travelling across a larger territory than ever before. This phenomenon is widespread and can reveal itself in many different ways; it includes kinetic nomadic elites travelling all around the world on high speed/high cost transports, as well as people who are obliged to travel, such as migrants. However, this phenomenon is also beginning to affect an ever growing number of ordinary people who travel along more mundane paths in their everyday lives. To this regard, and with reference to the study of everyday mobilities, some authors have highlighted that people tend to be more mobile every day, making longer journeys for work, in order to avoid more demanding forms of mobility such as relocation or migration. In doing so, they are able to manage both work and family. Kaufmann highlighted the increase of more reversible practices of mobility (Kaufmann, 2002: 25) at the expense of less reversible ones. Keeping in contact with loved ones when spatially far; listening to a song that recalls nice moments and places; carrying pictures and memories; as well as returning home every night even when working hundreds of kilometres far from home, are all different ways of reducing the impact of distance on everyday life, in other words to experience more reversible mobility practices.

The main aim of the book is to study emerging mobility practices looking at two main aspects: firstly, by analysing how far they have changed the way people use and transform spaces in order to better fit their needs and requirements, considering that the spatial effects related to the increase of this mobility practices are still relatively unexplored despite being widespread and relevant for improving urban design and territorial management; secondly, by understanding the impacts of emerging mobility practices on the relation between people and territory.

This book is the results of a PhD research in urbanism elaborated at the Politecnico of Milan. The research is based on a theoretical background as well as on empirical results, collected and elaborated thanks to direct observation of 11 highly mobile people through interviews, travel-along and interactive maps.

## Innovative perspective on emerging mobility practices: theories and methodologies

The developments of communication technologies and the diffusion of 'miniaturised mobilities' (Elliott and Urry, 2010: 21), in particular laptops and smartphones, allow people to change their role anywhere (working at the café, or carrying on the management of family life while on the train). The speed of travelling and communication promotes the simultaneous inhabitation of numerous spaces and allows us to combine in everyday life spaces of activities that were once impossible to match (Vincent-Geslin and Ortar, 2012: 40). As a consequence, people can 'avoid having to choose between alternatives such as moving and taking up a new job, between a partner and living alone' (Kaufmann, 2002: 42).

In the last 20 years, the increasing importance of mobility in everyday life advocates a deeper reflection on the concept, to the point of being able to observe a mobility turn and announce a 'new mobilities paradigm' (Sheller and Urry, 2006). The new mobilities paradigm collects approaches from different disciplines, such as: sociology, urbanism, geography, anthropology and transport studies. It emphasises the idea of a world on the move in contrast to a more sedentary model, and intends social entities and societal structure as part of a system in movement.

More generally, the appearance of a new mobilities paradigm brings attention to the analysis of mobility as a system consisting of many differing mobilities, including corporal travels, physical movements of objects, imaginative, virtual and communicative travels (Urry, 2007: 47) and invites us to take into account the different scales of mobility 'from the small-scale bodily movements, through infrastructural and transport aided movements to global flows of finance or labour' (Cresswell, 2010b: 552); it calls for a renewed set of questions and methodologies to interpret the contemporary world; seeks a more interdisciplinary approach that combines spatial and social mobility, and which takes into account the reasons and the effects of movements on different life domains; lastly, it demands a renovation of mobile methods, from fixed and static to more experimental and on the move.

For what concerns this last point, traditional approaches to the analysis of mobility may be misleading for various reasons: they fail to consider the reasons that lie behind mobility and do not consider mobility as a complex phenomenon that links and interconnects different domains of a person's life; they are static, referring to a static idea of space and time. This research tests a mobile method based on the user experience. This method allows a multidisciplinary and multi-scalar approach to the subject. Thanks to this approach, it has been possible to avoid a deterministic definition of mobility spaces. As a matter of fact, the research proposes to move on from an interpretation of mobility environments as 'places where mobility flows interconnect' (Bertolini and Dijst, 2003: 28) to a consideration of mobility spaces as all the places touched upon by mobile people in their everyday lives.

From the operational point of view, defining and experimenting with this method together with the redefinition of mobility spaces is one of the most important results of this work. The identification of three different profiles describing how the relationship between people and territory has changed due to the transformation of mobility practices and the suggestion of mobility design and solutions that may match different needs is another interesting outcome of this research.

## Chapter overview

The book is organised as follows. The first chapter presents the theoretical background of this research introducing the field of mobility studies with particular attention to those mobility practices defined as reversible, and considering how far changes in mobility behaviours have influenced the redefinition of urban rhythms, and the relationship between people and territory. Chapter two introduces the methodology of research and describes the research tools. The rise of the emerging forms of mobility in Italy is also partially highlighted by statistical data. National conditions as well as the different urban contexts that fostered the development of emerging mobility practices will be the subject of chapter three, which represents an interlude between the theories that allowed this research to come at light and the empirical results. In chapter four, I will present the interviewees, introducing their mobility stories and the main questions related to the development of mobility behaviours. This chapter shows how the people observed in this research are useful points of reference for the transformation of mobility practices. Describing their stories becomes an essential tool in understanding actual changes in territorial relation and urban rhythms. Chapter five focuses on the consequences of the configuration and uses of space: as high mobile people strain boundaries between different spatial dimensions and emphasise rhythms and intensity of use. In this chapter, I will answer the following questions: in which way are spaces and services that are traditionally attached to mobility transformed in order to give an answer to new needs and uses? Are there other spaces involved in this process of transformation that can also be considered as mobility spaces? And what are their spatial characteristics?

Finally chapter six presents the main conclusions and findings. The use of a mobile method of research helped in analysing mobility towards a different perspective and opened up to numerous questions that may represent the subject of a future research agenda on mobility spaces: the management of administrative borders, citizenship rights, private-public spaces, all matters that may suggest the rise of spatial inequalities and injustice.

# 1 Mobility and the spatial perspective

## Introduction: mobilities beyond space and time

In recent years the availability of ICT and the increasing use of smartphones, tablets and faster means of transport have tended to nullify the impact of spatial distance on travellers. The transformation from door-to-door to place-to-place and finally person-to-person communities (Elliot and Urry, 2010: 48), changes in the configuration of the nuclear family (Widmer, 2006), the development of networks of distant personal relations (Larsen, Axhausen and Urry, 2006), the massive access of women to work and the diffusion of new forms of job contracts (Colleoni and Zajczyk, 2003: 40) are all factors that have contributed to a radical transformation of mobility practices. Lastly, the recent economic crisis has led, on one hand, to an expansion of job search basins, influencing the inclination to mobility (Ravalet, Vincent-Geslin and Dubois, 2014), and, on the other hand, to the configuration of new job contracts.

In their book *Mobilité Spatiale*, Michel Bassand and Marie-Claude Brulhardt (1980) suggested an approach to mobility that could embrace all its different meanings – social, spatial, job mobility, residential and cultural (Bassand and Brulhardt, 1983: 51) – and affirmed that spatial mobility is part of a system that needs to be studied as a whole. To them, mobility is a fundamental social process constituted of all the different types of movements. While considering mobility as a fundamental social process, they laid the basis for a definition of mobility that could be both spatial and social, and which would overcome the limitations of a fragmented approach.

This approach has been toughened, in recent years, through the formalisation of a research field in mobility studies and the configuration of a new mobilities paradigm. In the article 'The new mobilities paradigm', Mimi Sheller and John Urry (2006) depict the basis of this emergent paradigm, which had found its origins some time before following the release of John Urry's (2000) foundation text *Sociologies beyond Societies.*

In this book, Urry suggests a new approach to mobility, dismissing an idea of society 'as a uniform surface' (Urry, 2000: 3) and highlighting that social mobility literature has always considered mobility only as a vertical category (social mobility happens when people change their social status) without registering 'the geographical intersection of region, city and place, with the social categories of class, gender and ethnicity' (Urry, 2000: 3). Instead, he moves the attention to 'the diverse mobilities of people, objects, images, information and wastes' (Urry, 2000: 1) and to the analysis of 'the flows of people within, but especially beyond, the territories of each society', and to 'how these flows may relate to many different desires, for work, housing, leisure, religion, family relationships, criminal gain, asylum seeking and so on' (Urry, 2000: 3). The use of 'mobilities' instead of the grammatically correct singular 'mobility' emphasises the complexity of the phenomenon and the multiple meanings that must be assigned to this word (Urry, 2007: 8). Furthermore, it suggests the will to investigate all of these different mobilities as a whole avoiding the parcelling between different academic disciplines. These different mobilities are listed as follows: 'the corporeal travel of people for work, leisure, family life, pleasure, migration and escape; the physical movement of objects to producers, consumers and retailers; the imaginative travel; virtual travel; and communicative travel' (Urry, 2007: 47).

A favourite smell or a familiar song can make us feel at home, in the same way that the taste of a soggy *madeleine* on a tea cup transported Marcel Proust back in time to the places of his childhood (Proust, 1913), demonstrating that mobility also has to do with the emotional dimension, and includes personal sensitiveness, virtual and imaginary trips, as well as physical movements. All those different forms of mobility go far beyond the four categories traditionally considered in sociology studies: daily mobility, travel, residential mobility and migration. After explaining how far the new mobilities paradigm has changed the approach to the study of mobility as well as sketching its main criticalities, in the next paragraphs it will be shown that this new approach also allows us to recognise some emerging practices of mobility and to study their impact on space.

Indeed, the development of ICT allowed the hypermodern individual to inhabit a time space with multiple dimensions, and to sail between multiple places and times. Facing this growing variety of space-time dimensions, he chooses techniques that allow him to move as fast as possible in between different domains, trying to catch the ubiquity and the simultaneity that help him to keep together his scattered life (Ascher, 2005a: 53). This means that, thanks to new means of transport and communication, people can move simultaneously between different life domains where different dimensions of life flow at different speeds in different places, jumping from one dimension to the other. The speed of travelling and communication promotes the simultaneous inhabitation of numerous spaces and permits the combination in everyday life spaces of activities that were once impossible to match (Vincent-Geslin and Ortar, 2012: 40), encouraging the appearance of those mobilities practices that have been defined as reversible and permitting the amalgamation of work time and family time, public and private spaces, spaces for work and those designated to family life. This process engenders an individual time that is cyclical and simultaneous (Melucci in Pasqui, 2008: 80), while social time is linear and irreversible. The rhythm of everyday urban life is defined by the dialogue between these two different temporalities: linear social time and cyclical, chaotic individual time (Lefebvre, 2004).

Furthermore, as a consequence of augmented mobility, individuals can use different places in different ways, generating a disjunction between people and territories (Pasqui, 2008: 24). The opportunities offered by emerging mobilities practices multiply the territories to which each individual may relate.

## An interdisciplinary approach to mobility and space-time relation.

Mobility behaviours have been the subject of analysis in many disciplines such as transport engineering, geography, urban planning and regional sciences, which have focused especially on the mass movement of people, studying commuters' travelling patterns and matching supply of transport with demand of movement, without ever stating an approach that could integrate all the different contributions. Notwithstanding the call for a more interdisciplinary approach for the analysis of spatial mobility (Bassand and Brulhardt, 1983: 50), nowadays it is still possible to recognise a fragmentation of mobility studies. While social science, for example, does not usually take into account spatial movement as a factor of transformation in everyday life, and refers to social mobility only as a change of role and social status, a transport approach refers mostly to movement looking at its spatial dimension, without taking into account the social implications of mobility and the reasons that push people towards a mobile life or its consequences. As Kaufmann pointed out, transport studies have focused 'on fluid dynamics-based models' and developed their 'own definition of movement in physical space and flows of particles, individuals, cars, motorcycles, etc. … in the space' (Kaufmann, 2011: 25). At the same time, sociologists have looked at career paths, migration, residential mobility and issues of social segregation using the term mobility to intend a change in social level. One of the reasons for this split was the difficulty, in transport studies, of managing the space-time relation in the construction of traffic models, which are based 'on the assumption of temporal stability' (Kaufmann, 2011: 25).

The opacity of temporal dimension on studies regarding spatial mobility has also been highlighted by Alan Tarrius (2000). He believes that spatial approaches stressed the preponderance of accessibility and spatial patterns at the expense of a more complex analysis of different mobility temporalities and reduce time to a vague dimension: a mere temporal interval (Tarrius, 2000: 42). Analysis regarding people's mobility has definitely failed to consider a more holistic dimension of mobility and to connect its different meanings: movement through space for the geographer, mobility as a flow of transportation for the traffic engineer, and change in social position or role for sociologists. This vivisection and fragmentation 'became an obstacle for the understanding' (Kaufmann, 2011: 23) of the subject.

Moreover, as highlighted by Vincent Kaufmann, the fragmentation of the approaches led historically to a focus on the temporalities of the travels and the spaces in which mobilities took place (inside or outside the population basin). This has thus configured four main categories of mobility: daily mobility is exercised in short temporalities and inside the living area; residential mobility is also exercised inside the living area but involves long temporality of movement. Mobilities exercised outside of the living area are travel, on short temporalities, and migration, on long temporalities (Kaufmann, 2002: 2011). Even though the studies of these four categories fostered much progress regarding the analysis of mobility, they remained fragmented. This limitation is particularly visible considering that the developments of transport and ICT together with changes in the job market and massive access of women to work engendered forms of mobility that barely fit into these four categories. As said in the introduction to this chapter, the increasing importance of mobility in everyday life has led to the definition of a new mobilities paradigm.

### *The new mobilities paradigm: a moving concept between innovations and limits*

In spite of the importance of the new mobilities paradigm, some criticalities may be highlighted. Following a famous statement by Bruno Latour (2005) regarding actor-network-theory, Cresswell has stated that 'there are only three problems with the New Mobilities Paradigm and they are the words "New", "Mobilities" and "Paradigm"' (Cresswell, 2010a: 159).

The term 'paradigm' is derived by Kuhn (Urry, 2007: 18) and suggests the idea of 'a normal science being transformed by sudden revolutions' (Cresswell, 2010a: 159). This is a dangerous approach as it can suggest that the whole world is on the move and notions linked to a more sedentarist approach such as border and place are no longer part of the contemporary world. Instead, as considered by Mimi Sheller (2011), the new mobilities paradigm emphasises both mobilities and immobilites and the relation between mobilities and moorings (Sheller, 2011: 5). Also the word 'new' may be misleading. When used to refer to 'mobilities' it may be easily rejected as different forms of mobility have always existed and people have always been mobile (Cresswell, 2010a; Vincent-Geslin and Ortar, 2012: 39).

Therefore, what are the advantages of talking of a new mobilities paradigm? As stated before, even within these limits the new mobilities paradigm has opened up reflections on spatial mobility and provoked attention to the concept. It casts light on the concept, inspiring new reflections and calling for a more interdisciplinary approach. Moving on from 'mobility' to 'mobilities', it points out the multiple ways of moving: corporal, physical, imaginative, virtual and communicative. It includes, at the same time, the analysis of the mobility of people, objects and ideas. And so it emphasises the dimension of complexity linked to the concept of mobility. In so doing it moves away from an interpretation of social analysis as 'a-mobile' (Sheller and Urry, 2006: 208), and from a sedentary understanding of social life that considers 'normal stability, meaning and place, and treats as abnormal distance, change and placelessness' (Sheller and Urry, 2006: 208).

Lately, definitions of mobility have circulated in accordance with the mobilities paradigm or against it, and many concepts and ideas about mobility have been defined, such as the 'right to mobility', as defined by F. Ascher (2005b), which affirms that the right to work, housing and studying are subjected to the right of being mobile. Thus the right of mobility has become a general right with a specific and increased importance for society: motility, which has been defined as the 'aptitude for movement' (Kaufmann, 2011: 37); network capital, which comprises 'access to communication technologies, affordable and well-connected transport, appropriate meeting places and caring significant others that offer their company and hospitality' (Larsen, Axhausen and Urry, 2006: 280); as well as some more recent interpretations of 'accessibility', based on capabilities approach, which define accessibility as the capacity that allows individuals to reach opportunities and resources in order to be part of a territory and a community (Lucas, 2012; Martens, 2017), or as a tool to enable potential mobility (Pucci and Vecchio, 2019). Thus, it is definitely possible to say that the mobilities paradigm has inspired a more interdisciplinary approach to mobility that brings together different scales of mobility and tries to overcome the boundaries between different disciplines, confronting the reasons behind the choice of being mobile with their translation on spatial movement, and thus returning the complexity of the phenomenon.

For instance, the new mobilities paradigm is, of course, a moving concept that has evolved consistently from its first formulation, also as a consequence of the critiques that many authors have made of it. Departing from a background based on metaphors such as nomadism and networks, and based on the configuration of 'space of flows' (Castells, 1996), many theorists of mobilities study nowadays agree that the 'current mobilities turn should not be confused with the use of those metaphors' (Sheller, 2011: 2); rather it must be considered more as a branch of knowledge that suggests 'questions, theories and methodologies rather than a totalising description of a contemporary world' that 'delineates the context in which both sedentary and nomadic accounts of the social world operates' (Sheller, 2011: 2).

## Between mobilities and moorings: reversible practices of mobility

The increase in mobility and the diffusion of fast means of transport have indeed produced two interconnected processes: first of all, 'people can travel, relocate, and migrate and still be connected with friends and family members "back home" and elsewhere' (Larsen, Axhausen and Urry, 2006: 261). This means that even when far from home and relatives, it is still possible to be connected with our personal relational network, as emerging mobility practices 'may make and contingently maintain social connections across varied and multiple distances' (Urry, 2007: 48), and thus may reduce the impact of forms of mobility practised on long temporalities on different aspects of people's lives. Then, fast transport allows us to cover longer distances, in shorter times than before, thus boosting forms of mobility such as long distance commuting on a daily basis and dual residence which are based on the preference for travelling more often or for longer periods, rather than relocate or migrate. These practices give people the chance to choose 'an environment based on the opportunities made possible by speed potentials and the environment's receptiveness to their projects' (Kaufmann, 2011: 3) and alter the relationship between space and time.

It is clearly important to note that these changes have been propelled also by major changes in society: after the 1970s, the massive access of women to work and the diffusion of new forms of job contracts (Colleoni and Zajczyk, 2003: 40) added an important element of pressure. These changes have been thoroughly analysed in Italy, configuring a time-oriented sector of studies (Bonfiglioli, 2001) based on the analysis of complex time (Zajczyk, 2000) and on the conciliation of different areas of life, particularly on women as caregivers. Furthermore, the transformation in the structure of family, as clearly reported by Eric Widmer (2006), has contributed in modifying mobility behaviours. While a few years ago the nuclear family was a privileged reference in family research, and the limit of the household was the only criterion in defining family members, from 1970 onwards the idea of other kinds of family began to emerge (Widmer, 2006). Such changes in family structures also influenced the configuration of personal relationships and the localisation of significant others. Lastly, and more recently, we also need to consider another factor of influence: the global economic crisis, which placed pressure on people to modify mobility behaviours regarding the dynamic of job searching. In Italy, for example, since 2007 so-called 'involuntary part-time'[1] jobs have increased, doubling in the last five years (IRES, 2013). More generally, it is possible to notice that unemployment risks play a central role on the inclination to mobility (Ravalet, Vincent-Geslin and Dubois, 2014). The data issued by the Job Mobilities and Family Lives research[2] highlighted the rise of people ready to move or commute over long durations more than ever due to the need for a job. Moreover, the data also highlighted that in crisis countries such as Spain, for example, high mobility allowed people to get out of a period of unemployment.

Definitely new mobility practices are the consequences of new opportunities as well as representing new constraints. As an unforeseen effect of the freedom of being mobile, people can choose to avoid relocation or migration and may prefer different forms of mobility, based on short temporalities such as commuting, which allow them to occupy more places at the same time, even though geographically distant, without renouncing their moorings. Nevertheless, while traditionally commuting has been exercised only inside an individual living area, nowadays it is possible to observe people commuting also outside their living area, thanks to the speed potentiality offered by new means of transport and communication. According to Deleuze and Guattari:

> if the nomad can be called the Deterritorialized par excellence, it is precisely because there is no reterritorialization afterward as with the migrant, or upon something else as with the sedentary (the sedentary person's relation with the Earth is mediatized by something else, a property regime, a State apparatus).
>
> (Deleuze and Guattari, 2010: 45)

Thus, Deleuze and Guattari consider that while the nomad builds his relationships with territory on the move, the migrant needs to reterritorialise. The forms of mobility presented in this book are neither nomadism nor a form of migration or sedentary life. Instead, as suggested by Mimi Sheller, they consist of the alternation of 'mobility and immobility' (Sheller, 2011: 4), the need to be mobile and the desire to maintain moorings, moments of speed and stillness. Those mobilities allow certain people to inhabit different places at the same time, and suggest new ways of building territorial relations, such as politopical dwelling (Stock, 2006), multilocalism (Duchêne-Lacroix, 2013) and multiterritoriality (Haesbaert da Costa, 2005). Vincent Kaufmann has defined them as 'reversible mobility' (Kaufmann 2002: 123).

## What territory for reversible mobility?

Describing the territory of mobile subjects means to describe and localise their activities by circumscribing their mobility practices, configuring what has been called individual functional space (Lévy, 1994: 241), or 'activity space' (Schönfelder and Axhausen, 2003), which is that part of the environment a traveller is using for his/her daily activities. Reversible mobility practices permit the management of different activities displaced over a vast territory, connected by transport and communication networks and by the individual practices of mobility. Reversible mobile people, reinterpreting and readjusting their spatial practices, are simultaneously dwelling in more places and defining with their activity space an extended territory that is a strategic construction (Crosta, 2003: 10), a result of the actions made by the individuals that inhabit it, a space that acquires meaning for its users thanks to the practices engaged upon it, and that is built by the different localisation of their activities[3] (Pasqui, 2008: 85). In so doing they transform the relationship with territory and build multiple connections.

Vincent-Geslin and Ortar refer to this phenomenon as reversible mobility witnessing a reversible use of territory and network (Vincent-Geslin and Ortar, 2012: 40). It is possible to assign two main meanings to the concept of reversible territory (Vendemmia, 2018: 267). Reversible mobile people, according to the cyclical rhythm of their movement, are spatially coming back to the starting point every time that a temporal cycle is ended, and thus we could say that their territory is spatially reversible. Nevertheless, while moving they go through linear social time and so even coming back to the point of departure does not mean to come back to the same space. The second way of interpreting the idea of reversibility is that reversible spaces are subject to uses and interpretations totally different and disconnected from one another (Bourdin, 2013: 185). Thus, looking at the territories built by reversible mobile people, it is possible to define them as reversible in the meaning that their delimitation, boundaries and uses are temporary, variable and ephemeral, defined by the activities that take place upon them in a given time, and can be redefined from time to time. According to this definition, the same place can assume different meanings for different people and be, at the same time, part of the territory of different mobile people.

## Conclusions

Traditional approaches to mobility have seldom taken into account the complexity of the problem and thus the research on the subject was fragmented. The new mobilities paradigm and other analysis in the field of mobility studies demand a more interdisciplinary approach that can look at mobility, taking into account all the different aspects.

In this chapter I have first of all outlined the theoretical framework of this research. Mobility has been defined as a fundamental social process (Bassand and Brulhardt, 1980), a 'socially produced motion' (Cresswell, 2006: 3) consisting of movement meaning and practices. In this sense: 1) mobility is a movement that takes place in a socially structured system and is thus performed differently and acquires different meanings from society to society. 2) The analysis of mobility practices enables an understanding of the society that has produced them. 3) Understanding changes in mobility allows us to analyse changes in time-space relations. Using this definition entails an interdisciplinary approach to the study of mobility, offers a dynamic view of spatial transformations and allows an observation of the change in urban rhythm and in the relationships between people and territory.

Then, focusing on the transformation occurred on mobility practices, I have highlighted the opportunities provided by developments in ICT and transport. People can 'avoid having to choose between alternatives such as moving and taking up a new job, between a partner and living alone' (Kaufmann, 2002: 42), preferring forms of mobility practised on short temporalities and long distances. Those forms of mobility have been defined as 'reversible mobile practices' (Kaufmann, 2002: 24), that is all those mobilities that, thanks to accelerated speed potentiality, allow us to combine in everyday life places of activities that were, before, spatially impossible to match (Vincent-Geslin and Ortar, 2012: 40).

In the last part of the chapter I highlighted how the appearance of reversible mobility had impacted on the relationship between people and territory. Reversible mobile practices allow users to temporarily, but cyclically, occupy a space. These routines favour the construction of relationships with multiple spaces inhabited simultaneously or crossed every day, and define fleeting territories intermittently activated and shaped by the practices of their users. This is the reason why those mobilities result as an interesting lens of observation of changes in the relation between people and territory.

## Notes

1 In Italy these types of contract are called 'contratto di solidarietà'. They are job contracts that forecast a reduction of working time in order to reduce the impact of the financial crisis or facilitate new employment.
2 For more details on this research please refer to: https://www.bib.bund.de/EN/Research/Surveys/JobMob/Job-Mobilities-and-Family-Lives.html
3 This interpretation is based on Raffestin. According to Raffestin, the territory needs to be described towards the practices that are performed on it (Raffestin, 1986: 92)

# 2 Research method

## The reasons for new mobile methods

The new mobilities paradigm (Sheller and Urry, 2006) argues for the testing of different methods related to mobility research. By approaching mobility from the point of view of the 'mobilities turn' (Urry, 2000), new 'kinds of researchable entities' (Büscher and Urry, 2009) are opened up to involve all the different components and systems of mobility in the analysis. Traditional approaches to the analysis of mobility may be misleading, and consider only a few aspects of the process, without taking into account its multi-dimensionality as they use data that refers to administrative entities and, in so doing, they may miss some dynamics generated by mobility itself. People moving between administrative borders create new spatial configurations that cannot be ignored while working on mobility. Moreover, traditional statistical data, though essential in defining urban population flows at the macro-scale, does not give information about the reasons for mobility (Colleoni and Zajczyk, 2003).

Urry and Law also reported the inadequacy of research methods that 'deal poorly with the fleeting' (Law and Urry, 2004: 403). In order to overcome the static nature of the traditional research approach, a mobile research method should attempt to track people's movements, translating people's paths on maps and recording the entire experience of mobility in order to get as much information as possible. This approach includes dialogue with informants in order to get direct information or commitment to an analytic agenda, that is 'to use empirical data to gain insight into some broader set of social phenomena than those provided by the data themselves' (Anderson, 2006: 387).

The new mobilities paradigm identifies seven different mobile methods (Sheller and Urry, 2006, 217; Urry, 2007: 39). Firstly, research on mobility needs to be related to 'patterning, timing and causation of face-to-face co-presence' (Sheller and Urry, 2006, 217); this means to understand the reasons that bring people together, where they go, which factors cause their movement, the places where they meet, and why. Secondly, there are methods that require a more active participation of the researcher (Laurier,

2010). Those methods foresee observation as well as participation in the phenomenon observed: field-notes, sketches, photographs or video recordings are used as tools for data collection. The third method is the collection of time-space diaries, whose main limitation is due to the fact that the object of the research has a highly active role in the collecting of data; this can sometimes lead to misinterpretations (Latham, 2003). Latham suggested transforming the diary into a performance, profiting from the diarist's 'narrative resources' (Latham, 2003: 2009). Based on these reflections we can affirm that diaries can be written, iconographic or recorded (Sheller and Urry, 2006). Sheller and Urry named other possible methods of mobility research such as cyber research, imaginative travel or mobility memory. They also consider the need to examine the 'place of in-between-ness' (Sheller and Urry, 2006: 219), which is part of the mobility process but immobile. Lastly, there are mobile positioning methods through GPS or telephone networks (Büscher, Urry and Witchger, 2011: 8). In particular, the use of ICT is rapidly changing the way that we are able to study urban mobility. Compared to survey data, the development of location based technologies (LBS) make it possible to collect bigger quantities of data with less effort and greater precision; nevertheless, the use of this data is still limited in Europe due to the EU regulation on privacy (Van Schaick and Van der Spek, 2008: 29; Ahas, 2011: 197). Much of the ongoing research, such as the analysis of phone data in the cities of Rome (Sevtsuk and Ratti, 2010) and Milan (Pucci and Manfredini, 2011), or on a national scale (Ahas, 2011), offer a glimpse at the potentialities of those methods.

## The research method

To answer the call for more mobile research methods this work has developed a qualitative approach based on semi-structured interviews, travel-along and mapping. This method has been inspired by recent research on mobility: the Job Mobilities and Family Lives research[1] has led to the construction of the semi-structured interview; other methods such as drive-alongs (Laurier, 2010), being 'mobile-with' (Bissel, 2009) and, more generally, the idea of a performed interview elaborated by Latham (2003) encouraged me to undertake a travel-along observation and to use maps as a material object that could make it easier for people to talk about their lives and spatial behaviours. The use of objects that could facilitate the interaction and make interviewees feel at ease was also confirmed by Stephanie Vincent-Geslin. The information collected during this process contributes to build biographies of mobility, made up of words, maps and pictures.

The method has been selected for a number of reasons: 1) Firstly, considering that people on the move represent a vector that connects different territorial scales and functions, this method may be helpful in overcoming some important problems linked generally to traditional urban design approaches and to mobility design in particular. These include the traditional separation between

*Figure 2.1* Travel-along highly mobile people

different disciplines (transport studies, urbanism and sociology, for example), or the division between different scales of reflections, functions and typology of building. 2) Secondly, this method allows the researcher to describe the spatial effects of mobility in a sensitive way and from the user's point of view, consequently avoiding a preset idea of mobility spaces. The spaces affected by everyday mobility are not only those with high flows of people, as for example the station or the airport, but all the spaces that are transformed to accommodate new uses and behaviours. This approach allows the researcher to skip administrative borders and to concentrate on the actual experience of mobility.

Finally, using an actor-based approach is useful to involve people into territorial analysis and not considering exclusively places (Hågerstrand, 1970: 7).

## The research design

In the following paragraphs I describe in details the research design. Using the snowball technique I selected 11 high mobile subjects, travelled alongside them, gave semi-structured interviews and asked them to draw their activities on a map. Thanks to these tools, I investigated the type of services used in the different spatial contexts; the geography of personal relationships; the different profiles of inhabited places; and lastly the tactics elaborated by the interviewees to deal with an extended space and the impact on spatial configurations. The small number of participants selected meant that I could afford to explore the rich details of their activities. The study was not designed to provide a complete survey of spatial effects due to changes in mobility practices, but to be interpretive rather than statistical.

### *The description of the sample*

I personally interviewed 11 mobile people; I travelled alongside eight of them, and asked them to draw interactive maps of their lives and personal activities. For each person I conducted a thorough investigation that, through interviews and direct observation, covered all the different domains of the interviewee's life and the spaces that he or she inhabits. The interview lasted approximately 50 minutes, while the travel-along varied from one to four hours and offered me the chance to observe people during their trips and to talk to them personally.

Interviewees are people who correspond to some basic requirements. In particular, they experience forms of high mobility on a daily or weekly basis. Furthermore, 'based on the opportunities made possible by speed potentials and the environment's receptiveness to their projects' (Kaufmann, 2011: 3), they inhabit various environments simultaneously. This is both a choice and a need, due to the opportunity to work in a specific sector, earn a better salary, permit both partners to work, live in better conditions, be surrounded by a more pleasant context, be closer to their partner, profit from the family network for child care or benefit from the proximity of friends.

Interviewees are all highly skilled workers, with good accessibility and inclination to mobility that exerts more reversible forms of mobility on different time-bases: they are long distance commuters, shuttles or double resident and relocated.[2] When I asked if they count the distance in time or kilometres, one of them answered: 'Time, also because if I had to think about how many kilometres I do every day, I'd be afraid: too many!' They have decided to be long distance commuters on a daily or weekly basis mainly because of unfavourable job conditions, or to temporarily relocate to a different city, keeping their sentimental relationship and administrative residence, in other words their main place of residence, elsewhere.

The choice of interviewees was taken based on some important literary findings. In particular I have considered forms of mobility able to connect different living areas on short temporalities, more than classifications such as mobility in/at work, to work and mobility because of work.

Although choice of mobility means was not a selection criterion, nine of the 11 interviewees use the train as their main means of transport, the remaining two use cars because of the bad quality of train connections between their home and workplace. The evidence contained in their interviews, however, has been useful in characterising the reasons behind their choice of mobility.

*Table 2.1* The interviewees

| *Name* | *Interview date* | *Mobility practice* | *Cities* | *Travel Along* | *Work* | *Birth date* | *Birth place* | *Mean of tran.* | *Children* | *Contract* |
|---|---|---|---|---|---|---|---|---|---|---|
| Maurizio | Sept. 13 | LDC | Tur/Mil | NO | Construc-tion | 1979 | Mon-calieri | HSR | NO | OE |
| Ettore | Oct. 13 | DR | Licc/Mil | NO | ICT | 1967 | La Spezia | Car | YES | OE |
| Roberto | Oct. 13 | DR | Anc/Mil | NO | ICT | 1974 | Naples | Car | YES | OE |
| Sandra | Jan. 14 | DR | Nap/ Rome | YES | Education | 1980 | Naples | TrainIC | NO | FT |
| Giorgio | Jan. 14 | DR | Rome/Mil | YES | Showbiz | 1971 | Siena | HSR | NO | Freelance |
| Beatrice | Oct. 13 | ON/DR | Tur/Mil | YES | Fashion | 1972 | Asti | HSR | NO | OE |
| Valentina | Oct. 13 | LDC | Tur/Mil | YES | Design | 1966 | Pavia | HSR | YES | OE |
| Alessia | Oct. 13 | LDC | Tur/Mil | YES | PR | 1982 | Turin | HSR | NO | Tempor-ary project |
| Gennaro | Sept. 13 | LDC | Tur/Mil | YES | PR | 1983 | F.Cesia | HSR | NO | Tempor-ary project |
| Veronica | Jan. 14 | ON/DR | Rome/Mil | YES | Com | 1976 | Naples | HSR | NO | Freelance |
| Lorenzo | Sept.13 | DR | Rome/Mil | YES | Education | 1971 | Milan | HSR | NO | FT |

Note: LDC = long distance commuter; DR = dual residence; ON = overnighter; OE = open ended; FT = fixed term.

### *The semi-structured interview*

The semi-structured interview was composed of six different sections. The structure of the interview was designed in order to catch the complexity of mobility management, and to investigate the reasons for mobility and the relationships that allow the development of a highly mobile lifestyle. It has a special focus on the mobility generated by job conditions; nevertheless, it covers different domains of life, with specific attention paid to spatial consequences generated by mobility practices. All the interviews were conducted in Italian and translated into English at a later stage.

*Table 2.2* The structure of the questionnaire

| | |
|---|---|
| | • Name<br>• Surname<br>• Telephone |
| **Origins and places** | |
| | • Personal bibliography<br>• Main place of residence |
| **Family life** | |
| | • Partner job profile<br>• Children, children care, grand-children, home and parents<br>• Housekeeping and relationship |
| **Job** | |
| | • Professional bibliography<br>• Current job situation |
| **Job mobility** | |
| | • Long distance commuting<br>• Overnighter<br>• Dual residence<br>• Relocation for job reasons<br>• Pluri-mobility<br>• Distant relationship<br>• Circumstances<br>• Consequences |
| **Personal characteristics** | |
| | • Ability to be mobile<br>• Health, stress and satisfaction |
| **Socio demographic** | |

The interviews were conducted by telephone. The interviewees fixed an appointment that was convenient for them. Many of them asked to be interviewed on the move, while coming back home after a working day, because, they affirm, 'this is the time I keep for myself, to do my stuff' or because 'this is time to waste anyway'; some others gave me an appointment during the weekend so that they could be relaxed and take their time to answer my questions. When on the move, the telephone interviews were sometimes interrupted because of problems with connection.

While recognising the importance of the results issued by the telephone interviews, and some of the strengths of this method such as the possibility of reaching a larger number of people and constructing a detailed database and the opportunity to compare the different results, it is also possible to highlight some limits. One of the main limits is to be found precisely in the use of the telephone itself. Though the interview is very deep and detailed, the telephone is an unfavourable means of communication with which to establish confidence between the researcher and the interviewee. Finally, the very nature of the interview clashed with the variety of the interviewees' everyday routines. During the travel-along, I attempted to overcome these limitations.

### *The travel-along*

Eight of the interviewees accepted my participation on a travel-along experience. The travel-along was a necessary complement of the semi-structured interview in order to have a better understanding of the observed phenomenon. The travel-along consists of following people during their most usual trips, observing them while practising mobilities, their personal relationships, their use of transport space, the people they travel with or meet during the trip, and finally the devices and, more generally, the objects that 'travel' along with them. Furthermore, I took advantage of this time to ask some more questions, and finally asked the interviewees to fill in the maps of their inhabited cities.

Travel-along stems from the need for a mobile method that could 'capture … the processes of movement as they happen' (Sheller 2011: 7), and follows recently emerged mobile methods such as 'walk-alongs, drive-alongs or being mobile-with' (Sheller 2011: 7). Going beyond a homogenised idea of mobility, such methods recognise that 'traveling is a way of making time' (Vannini, 2009: 11); travelling alongside the interviewees it is possible to observe and analyse all the different components of their travels and, furthermore, all the spaces touched upon while being mobile. As the interviewees were travelling by train, I could directly observe public spaces such as the train carriage or the train station, more rarely I was able to enter office spaces, though never into their houses. With regard to these last spaces, in order to understand in which way they change under the pressure of mobility practices, I used the data collected during the semi-structured interviews, the travelling interviews and the interactive mapping.

*Figure 2.2* Mapping high mobility

Thanks to travel-along, it has been possible to distinguish two large groups: the everyday commuters and the shuttles who travel only a few days per week. In the first case, they were travelling with a monthly ticket without seat reservations. I met the traveller in the station or on the train, after having received a text message indicating the carriage and seat. I was not always able to travel next to them because of the crowd. In the second case, the lower frequency of the trips allows the subject to reserve a seat before departure. In these cases I was able to reserve my ticket and make sure that I could travel next to them. I met them mostly on the train. In both cases once the trip had ended, depending on the availability of time and their personal attitudes, I had breakfast with them at the station or I followed them to the office. During the travel-along I took notes and pictures of the travellers, observing the objects thcy use when travelling, their strategies to feel comfortable in varying places such as the train station or the office, the people they meet along the trip, and the ones they keep in contact with when they are away. Monitoring people on their trip certainly helps to comprehend the process of being mobile in a clearer way.

### *The mapping*

Two different types of maps have been produced during this research. The first type was edited during the travel-along and is interactive. I asked the interviewees to fill in the maps of their inhabited cities taking notes of the activities they carry out in a normal week and of the places where they hang out. The second type of map is elaborated by the author and is a synthesis of the information contained in the interview and in the interactive maps. The first type of map is the representation of the simultaneous lives of the interviewees, and highlights the complex relationship taking place between the interviewee and the territory. It also reveals, in many cases, a functional and relational identification with multiple territories; the second type is more general and ranges from maps of life mobility, to understand the mobility biography of the interviewee and his or her mobile competences and inclination to mobility, to maps of the different spaces inhabited in a given time (a day, a week or a month); this is in order to highlight the coexistence of different territorial scales and the multi-temporalities of actions in everyday life. These last maps help to identify the relationship that mobile people establish with different territories.

Concerning the first type, I invited the interviewees to draw a map while travelling-along. It was an interactive operation that requested user participation; I provided touristic maps of the cities where the subjects live or work, post-it notes and markers and asked them to note the most commonly used spaces and activities. The choice of using a touristic map was taken in order to facilitate the positioning of people in the different places.

In this sense, the action of mapping is part of the analytical process and has a twofold significance. On one hand, it is a device that helps people to feel more comfortable during the interview and to remember many activities and details that had been overlooked in the oral description. On the other hand the map is a very rich and complete source of information to be analysed and read after the interview in order to complement the oral description.

Many of the interviewees were surprised by the results of the mapping process showing to what extent they are unaware of establishing relations with different territorial realities. One of them, for example, affirms:

> If you ask me where I live, I say Turin. Later, while I was putting my life on the map, I realised that all my activities are in Milan. There is another part of my life, which is the relationship with my partner: in Turin. So I can say that I live in Turin, because my loved one is there, but the rest, my personal interest: cinemas, theatres, exhibitions, are in Milan.

Furthermore, the action of mapping helped people to face spatial constraints and to better describe their conditions, discovering that they are sometimes more easily oriented in one place than another; additionally, they reveal important information about the way they inhabit the space.

Watching the interviewees drawing their lives on the maps was also an important part of the observation process: the way they get oriented in the city, their chosen means of transport, the points they use as references. The profusion of details that is part of the representation refers to the capacity of people to engage in multiples spaces in a given time and suggests a condition of ubiquity and simultaneity (Ascher, 2005a) that is investigated and represented in the second type of map.

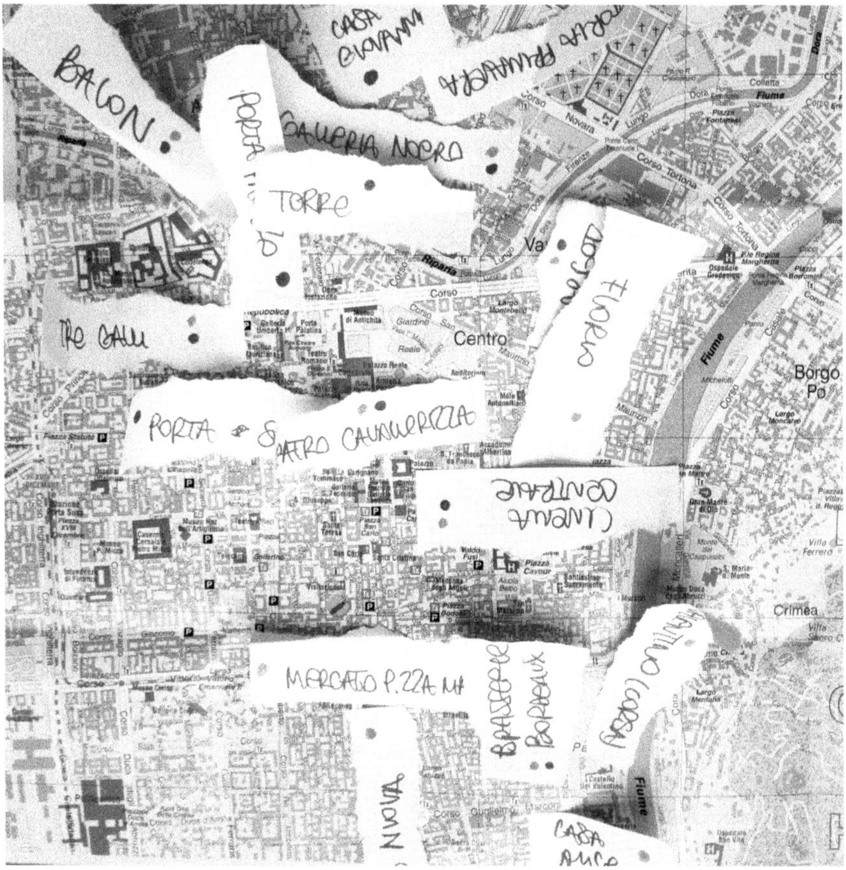


*Figure 2.3* Interactive map of Turin: Giorgio and Alessia

Representing biographies of highly mobile people is a great challenge for urban scholars, geographers and urban designers. This is because mobile people are 'vectors moving in many directions, switching positions and locations' (Sen and Johung, 2013: 3) crossing different scales over time. The representation of mobile geographies obliges urban thinkers to reflect upon the interconnection of space and time, changing scale again and again and understanding time not as a finishing line but as a dimension of people's lives. The interactive maps detected the 'ubiquity and simultaneity', without thoroughly representing the co-presence of different territorial realities. The cartography elaborated from the collected information represents, at the same time, different territorial scales: from international, national, interregional and city scales, including also that of the local neighbourhood, to the small scales of mobility places: it uses time dimension to keep together those different scales and to highlight 'topological relationships' (Lyster, 2013: 248).

These maps include maps of life mobility, an edited version of the maps of multiple cities and maps that compare the different spaces occupied in different time-lapses, trying to represent the so-called 'moment-place' (Pradel, 2015: 3; Vendemmia, 2018: 264).

The maps of life mobility represent the mobility history of each interviewee, highlighting how this influenced their way of being mobile and help to build up, or not, an inclination to mobility; furthermore, it shows different rhythms of mobility during workdays and at weekends.

The map of multiple cities focuses on a 5km radius zoom of the cities. At this scale it is possible to record the places frequented by the interviewees in each city, the activities, the paths that connect the different places, the means of transport chosen to travel from here to there, and finally the intensity of use and movement. The distances between the cities are represented in time. According to this representation, for example, Milan is as far from Pavia as it is Turin. Using kilometres the distance between Milan and Pavia is 44km, while Turin is 144km from Milan. This variance highlights that when changing metrics from time to space, a map is completely altered. Also the scale (Lévy, 1994: 69), cannot be fixed when mapping highly mobile people's lives. In order to represent the territory inhabited by the interviewees I needed to switch from the very large scale of interregional space to the very small scale of the train carriage, also including the instant connection to significant others.

Finally, a synthetic representation of activities and places shows the distribution of the different activities on a user-based territory, including activities on the move, and suggests important changes in the relationship between people and territory.

## A sensible approach to mobility spaces

Choosing a qualitative method based on the direct observation of individual mobility allows spaces to be analysed according to the experience people have when crossing or inhabiting them. According to Jacques Lévy, considering the scale of the single person is important because of the ability of each individual to perceive unexpected information and to enable unplanned encounters (Lévy, 1994: 49–50). In this sense, each of the interviewees operates as a sensor able to return feeling and sensation of a specific experience. For this reason, much attention has been paid to sensitive factors that rendered the spatial experiences of people involved in the research unique.

Furthermore, studying uses and perception of spaces helps to better formulate design questions. In fact, approaching spaces using the actor point of view permits us to look at all the different spaces involved in a mobile life such as, for example, house and office spaces, crossing different typologies and scales. In addition, with regard to traditional mobility spaces, an actor-based approach allows a better understanding of their real uses, highlighting where and why people choose to spend their time. Indeed, during the survey, while

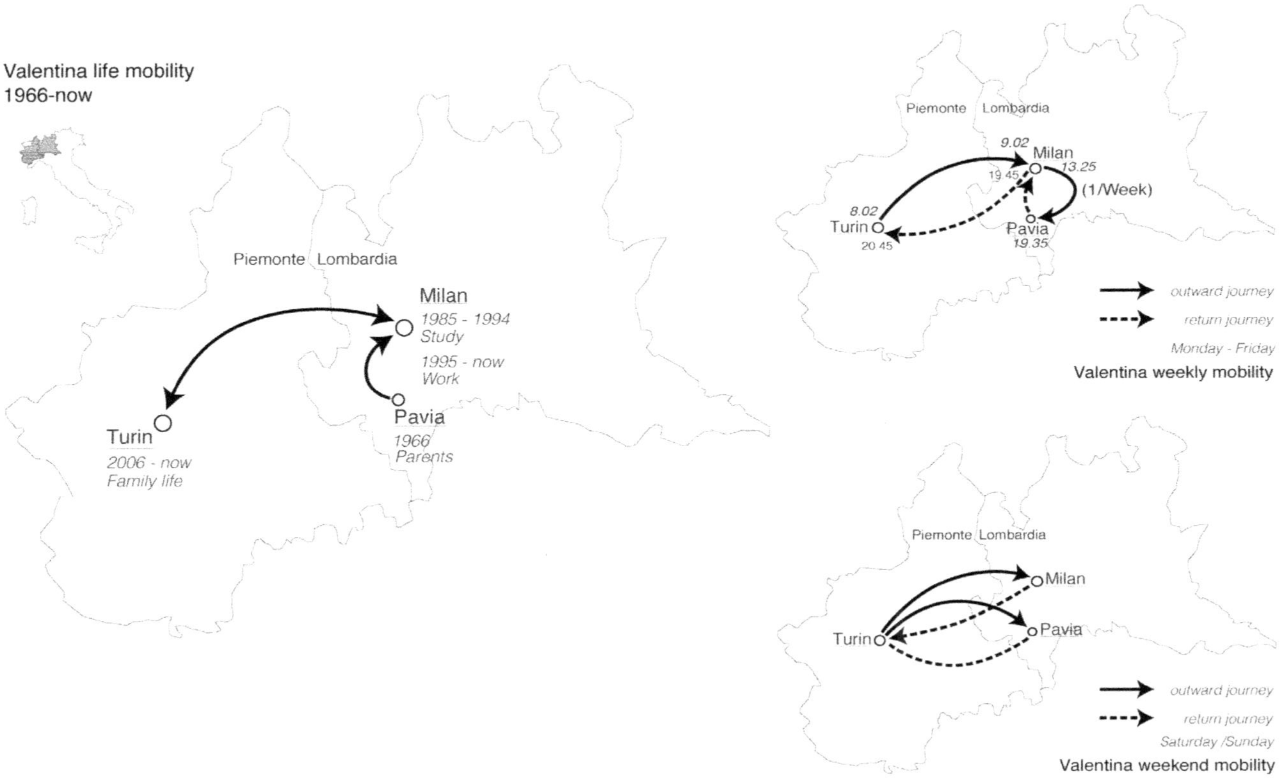


*Figure 2.4* Valentina's life mobility map

*Figure 2.5* Map of Valentina's everyday life

people were manifesting the need for services and spaces, they were also clarifying the main characteristics of those spaces requested in order to improve their quality of life.

Analysing mobility spaces through an actor-based approach follows a tradition started in France in the last decades of the previous century.[3] This approach highlights that mobility spaces are not only spaces of transit but also places for 'dwelltime' (Sheller and Urry, 2006: 220).

## Final considerations on methods

This research is based on a qualitative research method that departing from the direct observation of mobile people allows the analysis of spatial consequences of changes in mobility behaviours. I adopted this methodology because it allows three considerable problems associated with the analysis of mobility to be overcome. The first one is related to a traditional field approach to mobility. Due to the complexity of the subject, the use of a qualitative method of research can help to reveal its thickness and complexity, as looking at mobility from the point of view of the individual allows the inclusion and analysis of all the different aspects of life, such as work, familiar relationships, leisure and relational activities. It also reconnects the social point of view with the spatial one, joining temporal and spatial dimensions and covering different territorial scales. In addition, it gives the right importance to emerging mobility practices, even though they normally elude the toothcomb of statistical data, and allows the richness of the phenomenon to be highlighted.

Different tools such as interviews, travel-along and mapping permit different aspects of the problem to be covered. Semi-structured interviews allow, for example, an understanding of the change in social time, while maps help to understand the relationships established between people and territory. Moreover, their combined use helps to overcome the limitations of each tool: if the use of semi-structured interviews is too detached and not flexible enough to adapt to the different strategies elaborated by mobile subjects, travel-along offers the resolution to this problem. Furthermore, direct observation of mobility could reveal interesting insights about the relationships between people and their moving environment. Lastly, regarding space, this approach allows us to enlarge the reflection on mobility spaces to all the places encountered by a mobile life, avoiding an exclusive focus on traditional spaces of mobility, but highlighting new uses and meanings.

Nevertheless, I am aware that there are some limits linked to the use of this approach. Firstly, the results cannot be universally generalised. This is incidentally due to the fact that qualitative approaches are incisive when looking to explore a process in a very deep way, trying to catch its complexity; thus the results obtained are specific to the process analysed, and difficult to be generalised. Instead, the use of a quantitative approach can lead, if the sample is sufficient and variegated, to more generalising findings. Further limits are linked to the involvement of the researcher in the data collection and interpretation that could, in some cases, affect the subjects' responses, leading to improper interpretation of the results, or cause some important aspects to be neglected. This kind of analysis is also time-consuming both for data collection and interpretation and thus it can be costly. Furthermore, there is an issue linked to the anonymity and confidentiality of the data that can oblige the omission of some of the results. I believe that the use of mixed research methods that combine quantitative and qualitative approaches may help to overcome those limits.

## Notes

1 https://www.bib.bund.de/EN/Research/Surveys/JobMob/Job-Mobilities-and-Family-Lives.html
2 For this research, long distance commuters (LDC) are people who commute at least two hours every day, one hour there and one hour back. Considering that the majority of them commute by high speed train, they can be considered both long time/long distance travellers as they cover 140km every day; shuttles or double resident are people who are working and living in different cities, due to the long geographical distance between the two activities they decided to rent a house in the working place, where they live generally from Monday to Friday; finally relocated are mobile people who moved to another city for working reasons but still keep personal and family ties in the former location.
3 I refer here in particular to research driven by RATP in 1992 on intermodal hubs (Oudire, cahier RATP n. 76, 1992 in Pucci, 1996: 87).

# 3 Mobility in the Italian context

## Data and urban contexts

### Introduction

Analysing emerging mobility in Italy is interesting for at least three different aspects: first of all, because of the changes in mobility behaviours that have been registered also by mobility data on the national scale; secondly because of changes in social and economic structures, and last but not least because of the configuration of transport and the increase in high speed train connections. The national framework resulting from this socio-economic and mobility data exploration suggests that economic crisis and lack of job opportunities, together with the increase and speed of connections, fostered people's mobility towards more lively territories.

To better understand mobility dynamics in the country, in the last part of the chapter, I will introduce a portrait of the most important cities involved in the research according to some relevant indicators: number of inhabitants, dimension and availability of public spaces and garden, income and property values; all factors that may influence both residential and job locations. The descriptions will also pinpoint the different characteristics of the places: in some cases the availability of job opportunities and the liveliness of the context, in other cases the presence of low budget housing location, and local solidarity networks may represent important factors to attract new inhabitants. The data presented in this chapter corroborates the idea that changes in mobility practices depend upon numerous factors such as transformation of economic and social structure as well as technological innovation and infrastructure network, which both contribute to the establishing of new mobility behaviours.

### Changes in socio-economic structure

In 2011 a general census of the Italian population was elaborated by ISTAT.[1] This census helped to build a portrait of the socio-economic condition of the Italian population, registering among other factors the effects of the 2009 economic crisis. In particular the census documented an employment rate of 56.9%, which is lower than the European Union average (64.3%). In recent research on the effects of the financial crisis in Italy, it has been highlighted that Italian salaries suffered losses of up to 12% and that the poorest lost much more in income than the richest tenth of the population (OECD, 2014). The same report also considers that in 2013, youth unemployment in Italy exceeded 40%. The rate of NEET (not in employment, education or training) in Italy is higher than in Mexico or Spain. Young people frequently rely on family networks for support. The serious condition of unemployment, together with job insecurity, moved the age for moving out from the parent's house forward. Work-related instability also delays family formation, and Italy's birth rate remains far below the replacement level; from 2017 to 2018 there has been a birth loss of 18,000 newborns (ISTAT, 2018), generating depopulation especially in more remote and marginal areas. Furthermore, the number of marriages also decreased in 2011, emphasising the crisis of the traditional family structure. If on the one side, the age to be economically independent and to establish a family increases, also older people, already embedded in the job market, start to have problems because of the economic crisis. Since 2007 the so-called 'involuntary part-time' (work contracts that refer to people who are obliged by the employer to reduce their work time, or even not work for a period) has increased, having doubled in the last five years (IRES, 2013). All this conditions brought, on one side, an expansion of the job search basin; on the other side, they influenced the choice of house location sometimes forcing people to commute between the parent's house and the job location, or to establishing their residence place close to family and relatives to take advantage of parental help for child care, as will be confirmed by the results of qualitative surveys in this research.

The 2011 general census also offers interesting clues about mobility behaviours. According to this enquiry, in Italy, there are 29,000,000 people who commute every day for work or study, the equivalent of 48.6% of the resident Italian population. This population has increased by 2.1 million since the previous census in 2001. Two-thirds of these commuters are people who travel for work reasons. Furthermore, workers make longer journeys than students, who mainly travel inside the municipalities. Mobility time has also increased since 2001, as well as the number of people who travel more than 45 minutes every day. Regarding modal share, Italians continue to prefer travelling by car, although there has been an increase in train use along some specific routes (Ferrio, 2013). Residential mobility has also slightly increased.

This mobility trend is still growing as confirmed by the annual mobility report elaborated by the High Institute for Transport Education and Research (ISFORT, 2018), which highlights an increase of mobility rate since 2012 with a value of 88.5% in 2017, and reports that commuting for work reasons still represents 42% of suburban trips.

## The construction of HSR and the consequences on mobility behaviour

The Italian High Speed Railway (HSR) was conceived in 1990. The first part of the line between Florence and Rome was realised a long time before from 1977 to 1992 (Beria and Grimaldi, 2011) and the line has been active since 2009 even if some sections are still under construction. The HSR project includes both the implementation of the network as well as a general refurbishment and reorganisation of stations.

Concerning the connections and transport services, Trenitalia offers three different services to its Alta Velocità (High Speed) passengers: Frecciarossa has been active since December 2009 on the Turin–Milan–Naples–Salerno axis that accounts for 65% of the mobility demand;[2] Frecciargento trains travel from Venice to Rome and Naples, and also connect Verona, Brescia, Bolzano, Bari, Lecce, Lamezia Terme and Reggio Calabria; lastly, Frecciabianca is active between Venice–Milan, Turin–Milan–Venice, and Turin–Milan–Genoa.

Trenitalia, with Rete Ferroviaria Italiana (RFI), built the line as a legal monopoly (Beria and Grimaldi, 2011). Since 2012 the company Nuovo Trasporto Viaggiatori (NTV) has also been active on the same lines, and offers Italo train services between Turin and Salerno and from Venice to Bologna.

Since 2009 Italians have had the opportunity to choose a new and faster way of travelling, and it seems that the implementation of the HSR has encouraged more people to use trains. According to NTV, in the period from 1990 to 2009, the use of cars increased from 54% to 59%, flights from 11% to 20%, while the demand for travel by train decreased from 36% to 21%. Between 2009 and 2012, while the use of HSR train services increased from 39% to 54%, the use of cars decreased from 28% to 21% and the use of flights from 26 to 21% (Ferrio, 2013). Definitely the activation of the connections along the HSR in 2009 boosted the use of train at the expenses of other means of transport, both for time saving as well as for the increasing comfort of the travel (Viesti, 2019)

People travelling on HSR, according to NTV, are:

- new commuters on the HSR, that is to say people who have changed their city of residence, work or study thanks to the reduction of the travel time, and travel daily or weekly between home and the workplace;
- people who work or study far from the place they come from and, thanks to the new offer of transport, travel back home more often;
- and finally people who, thanks to the reduction in price (due to the competition provided by NTV), travel more for leisure.

Parallel to the redevelopment of the train line train stations have also been renewed by Grandi Stazioni. The train stations have been divided into four categories:[3] platinum, gold, silver and bronze. The 13 main railway stations along the line are managed by Grandi Stazioni with Ferrovie dello Stato

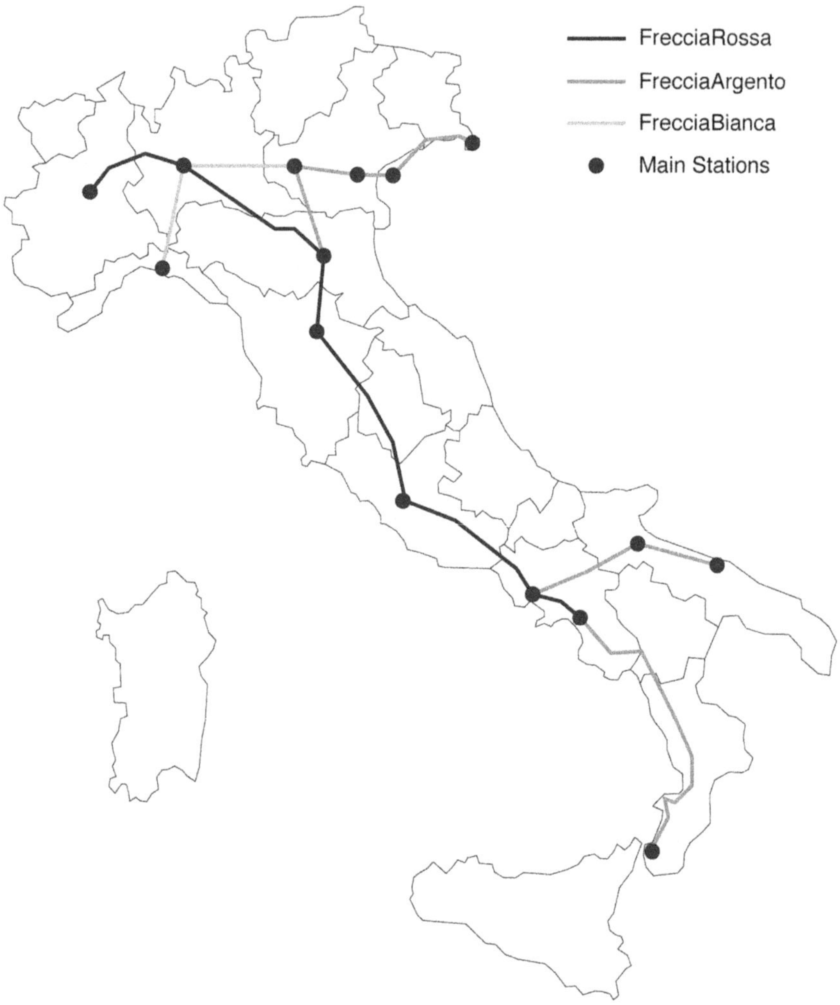


*Figure 3.1* The high speed railway in Italy

owning a 60% share; the other 103 main stations along the HSR line are managed by Centostazioni, though Ferrovie dello Stato is the major shareholder also in this case. RFI is responsible for the other stations (Moscarini, 2009: 224). The main railway stations also host Freccia clubs and Freccia lounges, which are the business lounges accessible only to Frecciarossa train clients holding gold or platinum CartaFreccia cards or to executive travellers. The train company describes these lounges as the ideal places to relax, work, read and be connected. It is also possible to invite one guest, thus they can often be used for small work meetings. Since 2012 some of the main stations have also held Casa Italo ticket offices.

Freccia clubs and Freccia lounges are located in Milan, Turin, Rome, Naples, Florence, Bologna, Padova, Verona and Venice. Milan, Rome and Turin are also the cities inhabited and crossed more often by the subjects analysed in this research.

## The cities involved in the research

Most of the interviewees reside in large urban agglomerations such as Turin, Milan, Rome and Naples, particularly in the city centre only one lives in the Neapolitan suburbs. During the interviews other cities have been cited: Fossacesia in Abruzzo, Asti in Piemonte, Pavia in Lombardia, Longarone in Veneto, Viterbo in Lazio, Siena in Tuscany. These are more commonly the cities where the interviewees come from and where their parents and relatives are still living. Although, notwithstanding those last cities are part of the mobility biographies, they are not part of the interviewees' activity space which 'is that part of the environment, which a traveller is using for his/her daily activities' (Schönfelder and Axhausen, 2003: 274).

In the following review I introduce some main considerations about the cities of Milan, Turin and Rome, which are the ones most present in the interviewees' everyday life.

The urban context can certainly have a significant influence on people's everyday lives and mobility practices. The same subjects can assume very different behaviours depending on the different places they inhabit and completely change their rhythms from frantic to calm; but also the same city may have different roles according to different moments of the interviewee's life: to the same person Milan may be the home town in a moment of life and become the city where commuting for work reasons in another moment.

### *Milan*

Milan is one of the cities that the interviewees visit most often for work reasons. This is also mainly because many of the interviewees work in sectors such as fashion or design. Milan hosts the world's most important design fairs, the design week, and together with London, Paris and New York is one of the world's four fashion capitals. The metropolitan area of Milan has a population of 3,088,791 inhabitants,[4] but the aforementioned characteristics make it a favourite destination for various people. There are city users who visit the city for different periods of time: for one day, one week or one month. Furthermore, Milan is well connected to other European and world capital cities due to infrastructural and geographical reasons: it is located in the north of the country, its three airports were respectively the 2nd, 3th and 4th for passenger traffic in Italy in 2014, as reported by Assaeroporti,[5] and it is home to two important train stations (Centrale and Garibaldi) that provide national and international connections.

*Figure 3.2* The cities involved in the research

Of the 11 people interviewed in this research, nine work in Milan. Four commute every day and generally inhabit the city from Monday to Friday, from 9 A.M. to 7 P.M.; four are shuttles who arrive on Monday morning and depart on Friday night; only one lives in Milan but spends his weekends in Rome.

Milan constantly offers new activities, restaurants, cultural events and shops and this is considered by many people as an attractive side to the city, although Milan property prices are among the highest in Italy: around 4000 €/sqm on sale and 20 €/sqm rental,[6] which make living in the city very expensive.

Another important consideration must be made regarding the way in which people move around in the different cities. Generally, Milan is considered in Italy to have a very good public transport system, as confirmed by the density of local public transport services as well as by the offer of passage/day (Beria, 2018: 126). The public transport services in the city of Milan are vast and varying and include also services of sharing mobility. Furthermore, Milan has the most car sharing users in Italy (Beria, 2018: 148)

It is the vast array of things to see and do at different times of the day, the abundance of big events and an efficient system of transport that make Milan such a vivid place. These factors allow different people to live in the city according to their temporal needs, and it is the overlapping of these multiple vibrations that create the rhythm of the city.

### *Turin*

Turin is the capital city of the Piemonte region; its metropolitan area counts approximately 2,300,000 inhabitants. Five of the 11 interviewees live in Turin though not one of them works there. Of the five interviewees that live there, only one is native to the city. There are various attractions to living here: the cost of living, the housing market, the dimension and the services offered, including public transport and sharing mobility, but also personal and familial relationships. In fact, the housing market in Turin is considerably cheaper than in Milan. The price of an apartment in central Turin may cost between 2750 and 4000 €/sqm, with an average price inside the municipality of 1700 €/sqm less than the half of Milan, where the same apartment may cost between 5400 and 7800 €/sqm.[7] These factors make Turin particularly attractive as a housing location. Apart from economic reasons, such as a lower cost of living and cheaper house prices, Turin is attractive to many people because of its human and familiar dimensions that make people feel at ease and facilitate the development of a solidarity network between family members and friends, and the enhancement of personal relationships and friendship. Furthermore, Turin offers a pleasant natural environment with 14.6% of the total surface covered by green areas (ISTAT, 2013a).

### *Rome*

While Rome is the most expensive city in Italy, it is still attractive for many people especially for the job opportunities found there due to the presence of national television headquarters, embassies, administrative centres, universities and research centres.

Covering a surface of 1287.4 km$^2$, the municipality of Rome is the biggest in Italy, and represents 0.4% of all national territory (ISTAT, 2013b). However, its territory is served by only three subway lines. The lack of public services and infrastructure, and the inconvenience due to the dimensions and morphological configuration of the city are counterbalanced by the fact that

Rome offers a very beautiful environment and many historical gardens. Rome has 45,600,000 sqm of green areas, of which 28,000,000 are historical gardens, thus making Rome the second Italian city for historical green surfaces (ISTAT, 2013a). The housing market is quite expensive with an average cost of housing properties rating almost 3000 €/sqm, with a price range between 2000 €/sqm in the suburbs and more then 6000 €/sqm in the city centre. In spite of this, Rome is appealing for the astonishing beauty of its historical centre and public spaces.

## Conclusions

The observation of socio-economic and mobility indicators on the national scale may suggest some important changes in mobility behaviours, such as, for example, the relationship between the insecurity of job conditions and increasing inclination to mobility, linked both to the need for travelling to reach better job conditions and the choice of housing locations due to familial and economic constraints. Long distance mobility to the workplace has been supported also by the development of high speed railway lines and made more comfortable thanks to the design of more pleasant mobility spaces. It is clear that the data presented here are only suggesting in a blurry and confusing way the emergence of new mobility practices. The portraits of the cities that are more cited in the research, according to some variable such as cost of living and property prices, may suggest some of the reasons that bring people to be mobile, highlighting for each city the most and less attractive aspects.

However, the data do not allow us to understand and describe the reason beyond changes in mobility behaviours, neither may they suggest how those practices affect the relationship between people and territory, showing that a qualitative analysis of the phenomenon is needed.

## Notes

1 ISTAT is the Italian National Statistical Institute (Istituto di Statistica Nazionale).
2 www.rfi.it
3 http://www.rfi.it/rfi/LINEE-STAZIONI-TERRITORIO/Le-stazioni
4 Data are for Milan Metropolitan area. http://demo.istat.it/bilmens2013gen/index02.html.
5 http://assaeroporti.com/statistiche_201410/
6 https://www.immobiliare.it/mercato-immobiliare/lombardia/milano/
7 https://www.immobiliare.it/mercato-immobiliare/piemonte/

# 4 Reversible mobility the length and breadth of Italy

## Redefining rhythms and territories

### Introduction

The subjects interviewed in this project were selected using the snowball technique and according to three main characteristics. Despite the homogeneity of the sample, the reasons behind their mobility choices are many and diverse.

The case studies analysed in this research are not representative from a statistical point of view, but the detailed description obtained during the interview and travel-along allows a better understanding of some processes caused by changes in mobility practices.

The interviewees were selected according to the following characteristics: 1) they all experience emerging mobility practices and they are long distance commuters (LDCs), shuttles and overnighters. 'It is a completely different dimension compared to a normal home-work schedule,' says Gennaro to this regard. 2) Their practices of mobility began mainly for reasons related to work, but have extended their influence over many different domains: in particular over personal and family life. 3) Due to their mobility practices they all simultaneously occupy various and diverse living areas.

On the basis of the thought process in deciding to be mobile, it will be possible to sketch some profiles: moving because of job insecurity, chasing better work conditions with family constraints, struggling between professional aspirations and personal ties, and finally being in a long distance relationship.

Furthermore the interviews allow us to underline that territorial relationships are changing in that they are becoming less sedentary and more dynamic; at the same time this transformation brings new city users, changing the schedule of uses and creating moments of high and low activity intensities. People observed in this research are pioneers of this emerging mobility behaviours in that they are useful points of reference of the phenomena. Describing their mobility becomes an important tool in describing actual changes in cities and in defining the spaces of mobility.

### *The main characteristics of the interviewees*

The interviewees differ in sex, marital status, number of children and condition of mobility. In particular they are 11 people aged between 31 and 48 years old, of whom six are men and five are women. Nine of the 11 are in a couple, of these three are married and six are in a common law marriage, four have children (three married and one in a common law marriage), while five are in couples without children. Four of the interviewees are LDCs; six are double residents, one of whom is also an overnighter; the last one is relocated and an overnighter, and also in a long distance relationship. When travelling by train, they travel by HSR. Beatrice says that the HSR has completely changed travel conditions, and she affirms 'we are lucky because we travel in HSR'. The same was said by Giorgio: 'I am a lucky man, I can travel on Frecciarossa, I don't have any disadvantages of travelling and I have an iPhone so I'm always connected', confirming that different forms of mobility are always interconnected. They work in creative sectors (visual merchandiser, costume designer, designer), communication (press agent, reporter), ICT and education; they are all highly qualified with bachelor, master or PhD degrees, and they have different job contracts: five have an open-ended contract, two are freelance professionals, one has a fixed term contract and three are on temporary projects. It is worth noting that only one of the women interviewed has children, while three of the men do. Their mobility behaviour is reflected in the immobility of their wives who are in charge of the children and the management of the household.

Their economic conditions are also different: they are neither rich nor poor. They do not correspond to the description of high mobile elite, travelling by plane all over the world in business class, nor do they travel by regional train (which are the slowest, least comfortable and cheapest trains in the Italian railway system). Their mobility is both a sufferance and a choice. A sufferance because it is, in most cases, the only way to get a well-paid job, for both partners to work, or to earn more money and enhance the opportunity for career development; it is a choice because, of the varying opportunities available, they have opted for a more intense daily, or weekly, mobility. To this regard Alessia, who chose to be a LDC between Turin and Milan, says:

> I was a bit down when I had to start working in Milan. It was a difficult decision, because I didn't want to live in Milan, but it was my choice to take the train. And at the beginning working in Milan was very tiring. The idea that I had to travel so long to go to work every morning and to come back was not even imaginable for me. I thought it was something temporary and I would soon have a job in Turin.

Because she cannot find a job in Turin, the city where she comes from, Alessia has decided to be a long distance commuter. She thought it would only be temporary but since then four years have passed by. She decided to take the train, the high speed line between Milan and Turin had not been completed at that time, and so her trip was quite long. When the line was finally implemented, she saved 40 minutes on each part of the journey, and so continued to commute each day. The interviews were conducted in February 2013 when the consequences of the 2009 financial crisis in Italy were visible in the uncertainty and insecurity of job conditions.

The interviewees are: Beatrice, Gennaro, Valentina, Giorgio, Maurizio, Veronica, Lorenzo, Roberto, Ettore, Alessia and Sandra. Roberto, Ettore and Maurizio did not want to travel-along although the information collected during the interviews helped to elaborate some reflections especially on the transformation of urban rhythm. The information collected during the travel-along contributed to building the different profile of mobility as well as the maps of activity spaces and the territorial profiles. The first names used to identify the interviewees are not real. This choice has been made in order to preserve their anonymity and, at the same time, to freely use all the data.

## Different profiles behind being mobile: job opportunities, economic conditions and family constraints

The choice, or the need, to inhabit one or more cities is based upon a complex system of variables in most cases.

Looking at the case studies, the proximity to family is, for example, one of the elements that most influences residential choices. A proximity to family allows children to grow up in a familiar context, and in many cases offers significant assistance in the management of family and personal duties.

Another important factor of choice is represented by personal preferences. Seven of the interviewees affirm that one of the advantages of being mobile is that they can live in a more attractive city. Maurizio and his partner decided to live in Turin because ‘it is our city, the place where we feel at home’.

An interesting insight is also related to the choice of neighbourhood, which determines the possibility to access different services such as shopping, mobility, schools. One of the interviewees declares:

> my house is well connected by public transport, so it’s not important if the central station is really close, but rather if I can easily reach it. If we could find a house in a different neighbourhood that was equally well connected, then for me it would be the same.

This suggests that being well connected is more important than being very close from a geographical point of view (Amin and Thrift, 2002), but the attractivity of the city has more impact on residential choice then the location of the neighbourhood.

Good accessibility to public transport is an important matter also for dual residents such as Beatrice, Giorgio and Veronica. Almost all of the interviewees live in the city centre though this does not seem to be a discriminatory factor. The Job Mobilities and Family Lives research demonstrates that the diffusion of long distance commuting is a common phenomenon even for people who live in suburban areas as in the case of Sandra, for example.

In some other cases we find circumstances that encourage the making of a choice: when Alessia found her job in Milan she had a relationship in Turin, her partner was living and working there, and so she decided to be an LDC instead of relocating to Milan.

Undoubtedly income plays an important role in mobility choice. While Beatrice can afford to rent two houses, Sandra would prefer to relocate to Rome but cannot afford the monthly rent. As a consequence, she stays in Rome only when she must and takes advantage of friends' hospitality or stays in a no-frills hotel.

### *Sandra – a shuttle because of job insecurity*

Sandra lives in Caivano, a small town in the suburbs of Naples, with her parents. She is very highly qualified and has a PhD degree in Environmental Architecture, though it has been very difficult for her to find a job in her field of studies. At the moment she is working as a substitute teacher in Rome. These extremely uncertain working conditions also make her life extremely mobile. Currently, she is a double resident of Caivano and Rome.

Following her PhD, Sandra moved to Bologna for one year where she worked as a technician for the army. The army gave her a good salary and accommodation in Bologna, but the job was temporary. Therefore, after eight months she went back to Caivano, to live with her family once again, and began the search for another job.

After a few months she had the opportunity to work as a substitute teacher in Rome, on a one year contract. She would love to move to Rome but:

> the salary I get for this job isn't enough to rent a house in Rome, so three days a week I stay at some friend's house or in a hotel. I start work at 8 o'clock in a neighbourhood 20 minutes by train from Rome central station, and leaving from Naples every morning would be too much.

Sandra would prefer to live in Rome but she is unable to do so because of the insecurity of her job and the meagreness of her salary. As she cannot live in Rome – 'I can't do it now, it's not my choice. I would love to live in Rome and be less mobile in my everyday life but I can't afford it' - but refuses to commute everyday – 'I wouldn't like to be an LDC as I wouldn't enjoy the time I spent in Rome, I'd only spend more time on the train. I know many people, colleagues that go and come back every day from Rome to Naples' – she has decided to be a shuttle so that 'when I am in Rome I can also meet people and spend time with friends, go to the theatre, visit exhibitions, I use the time I am in Rome for my relational life and for cultural activities, or leisure'. She was also able to make this decision as she is not in a stable relationship nor does she have children.

It is clear that Sandra did not choose her mobility. Rather, it is defined by her continuous search for a job, by the salary she receives from time to time, by its instability and uncertainty, and by economic conditions.

### *Giorgio – a nomadic job*

Giorgio was born in Arezzo. He is a freelance costume designer and counts many Italian TV channels among his clients. Giorgio is mobile because his job moves. 'I work in show business, we are like gypsies, how do you think they can live without travelling? My work is always on the move, it's not a static job,' he says. Due to the fact that he is freelance, and he follows different programmes, he moves continuously between the many studios located in Milan and Rome. In Milan he works during the day from Monday to Friday afternoon, but on Friday night he travels to Rome to follow another show. 'When I arrive in Roma Termini I take a taxi and I go home. If I have to work, I take the scooter and go to work.'

Giorgio likes to be mobile – 'I like to travel, I think it's part of my lifestyle' – but this is also because he chose his job and his mobility is a direct consequence. Furthermore, by being more mobile he can earn more money and have better job opportunities. His clients refund his costs of transport, and his journeys are comfortable. In addition, the context of his life is mobile as his colleagues are also always on the move.

Giorgio's mobility is a necessity linked to his job. Mobile jobs have always existed, and it is the ease and speed of Giorgio's trips that make the difference and gives him the chance to work in various places, managing his work and his private life over a large territory.

### *Maurizio and Valentina – chasing fixed job and higher salary with family constraints*

Maurizio moved to Turin in 2010 and had a child in 2014. He works in Milan and is an LDC. Valentina was born in Pavia, and is an LDC between Milan and Turin. Milan is not only her place of work, however, as she has lived there for 20 years.

Valentina and Maurizio's stories and decisions of mobility are quite similar. They have both been living in Milan for a while and when they started families they decided to move to another city. This city could offer them both better opportunities of finding child care, mainly due to the support of their parents or families, a more affordable house market and the possibility of a more comfortable urban environment. At the same time Valentina and Maurizio have open-ended contracts and would not find the same job conditions in Turin, nor could they work in the same sector. Thus they are LDCs struggling with their careers and professional opportunities, on one hand, and family life, on the other. They need to work to have a well-paid job and so being mobile is both an opportunity and, at the same time, a necessity.

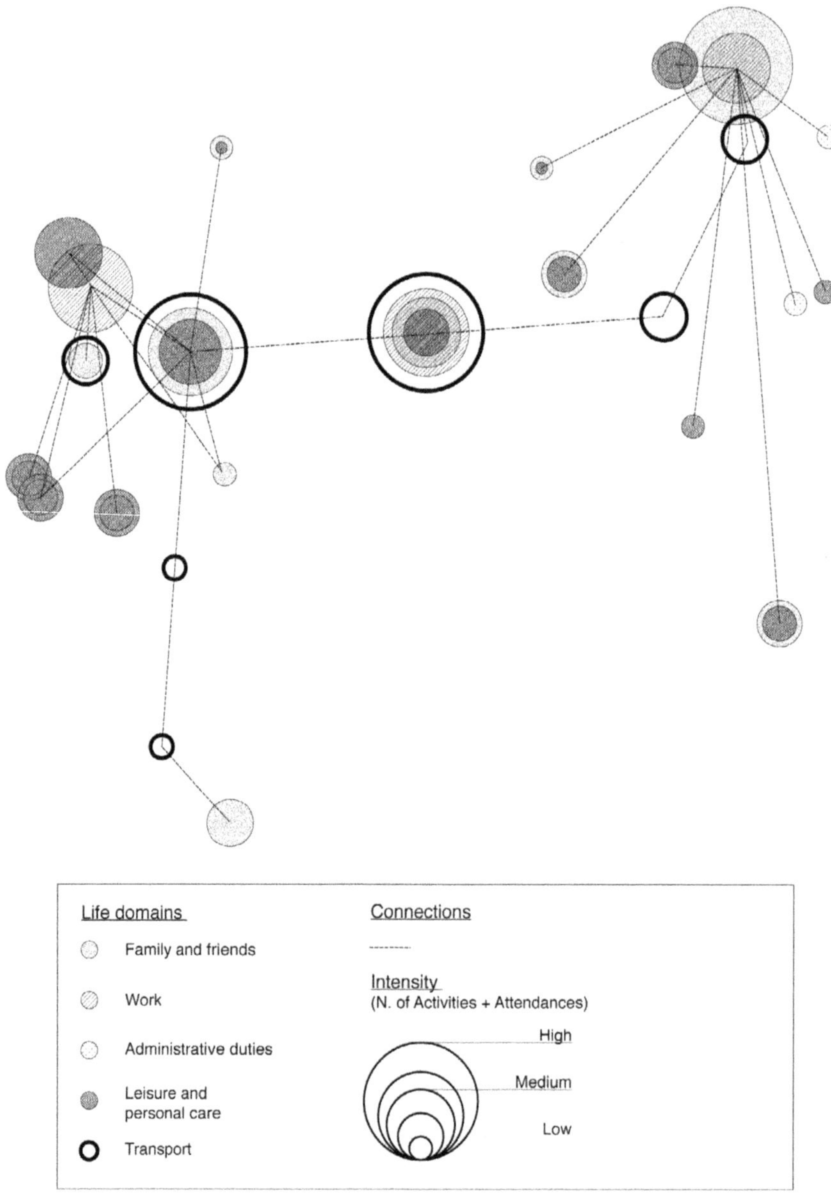


*Figure 4.1* Valentina's map of activities space

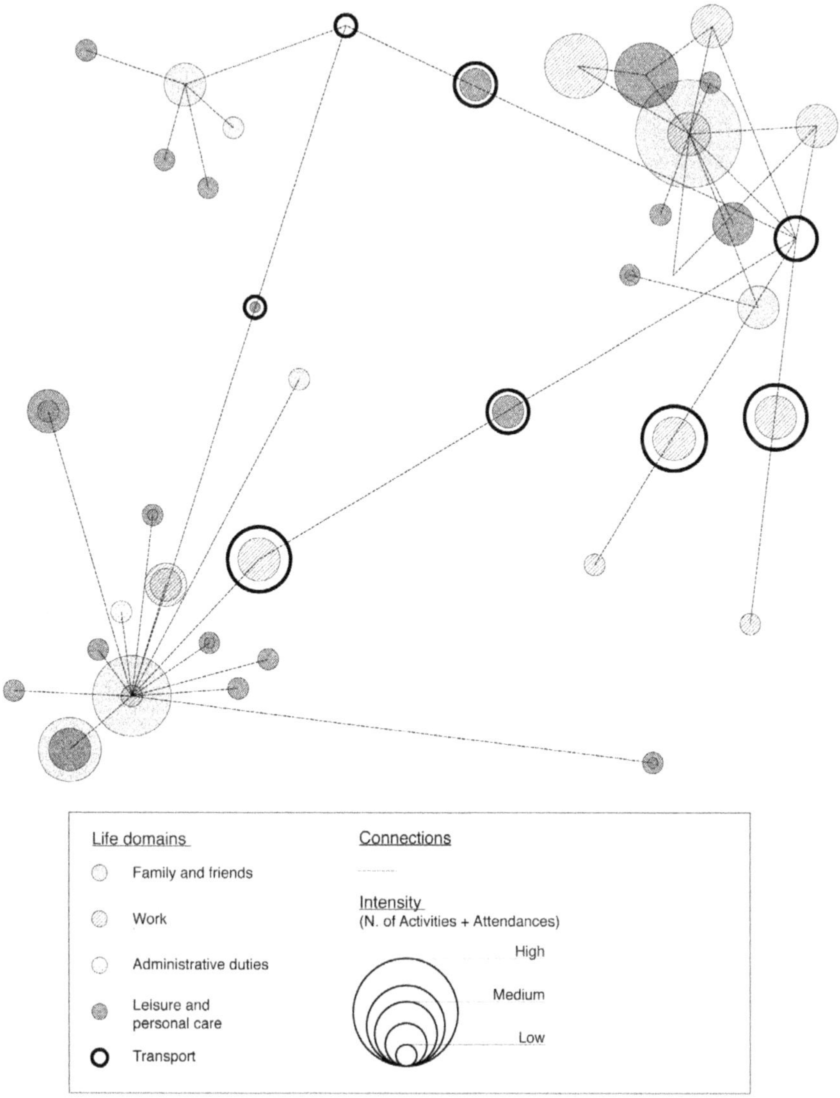


*Figure 4.2* Veronica's map of activities space

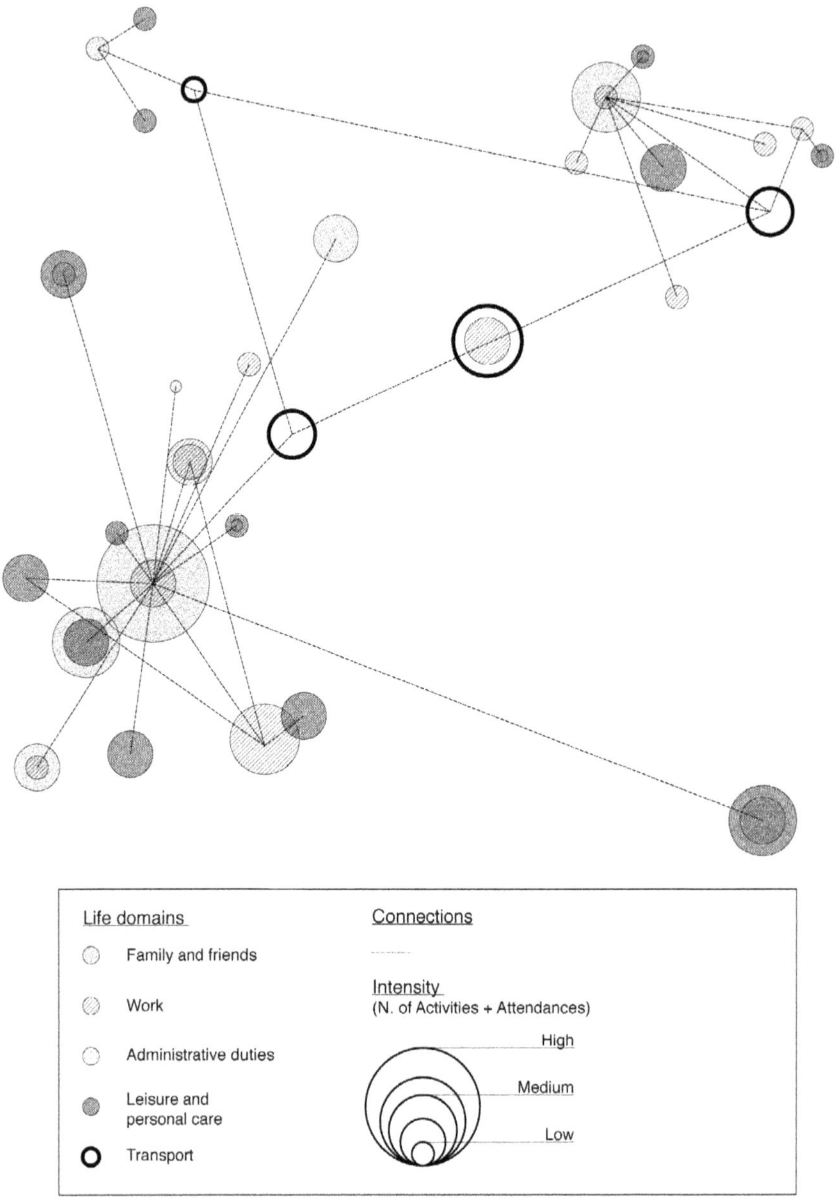


*Figure 4.3* Lorenzo's map of activities space

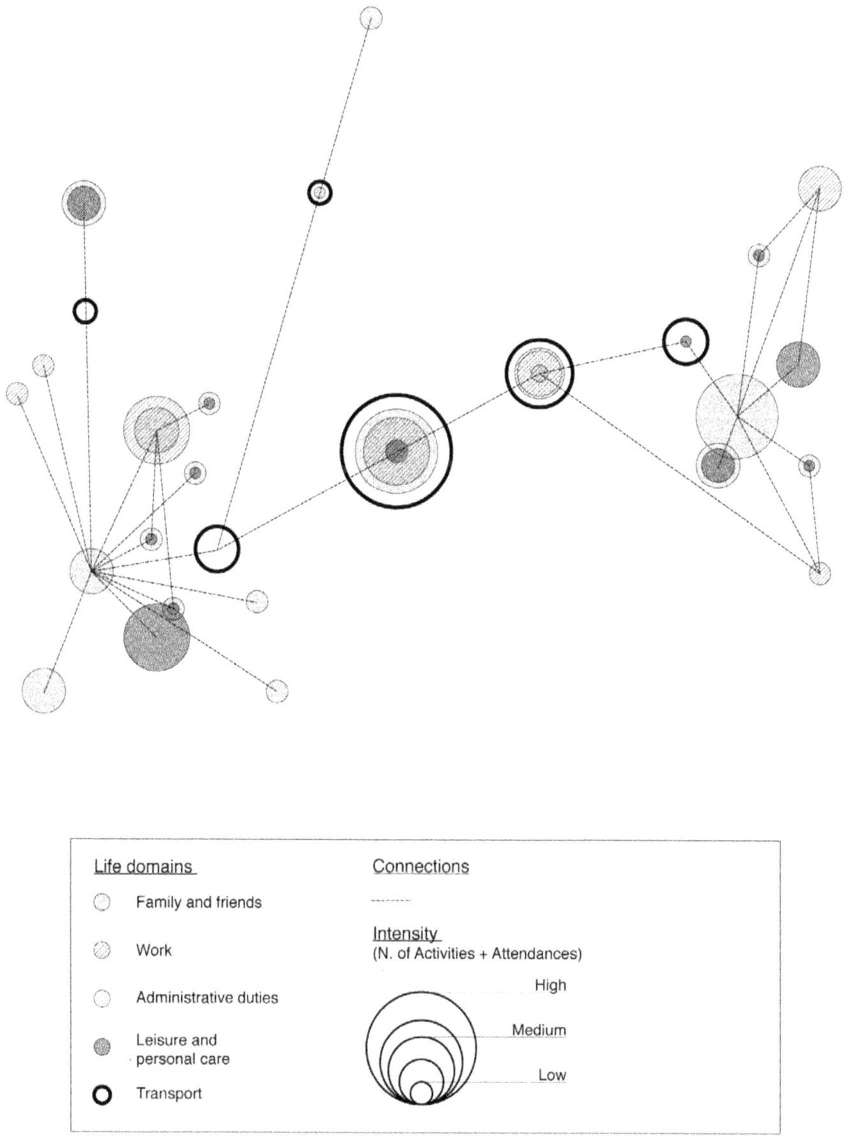


*Figure 4.4* Giorgio's map of activities space

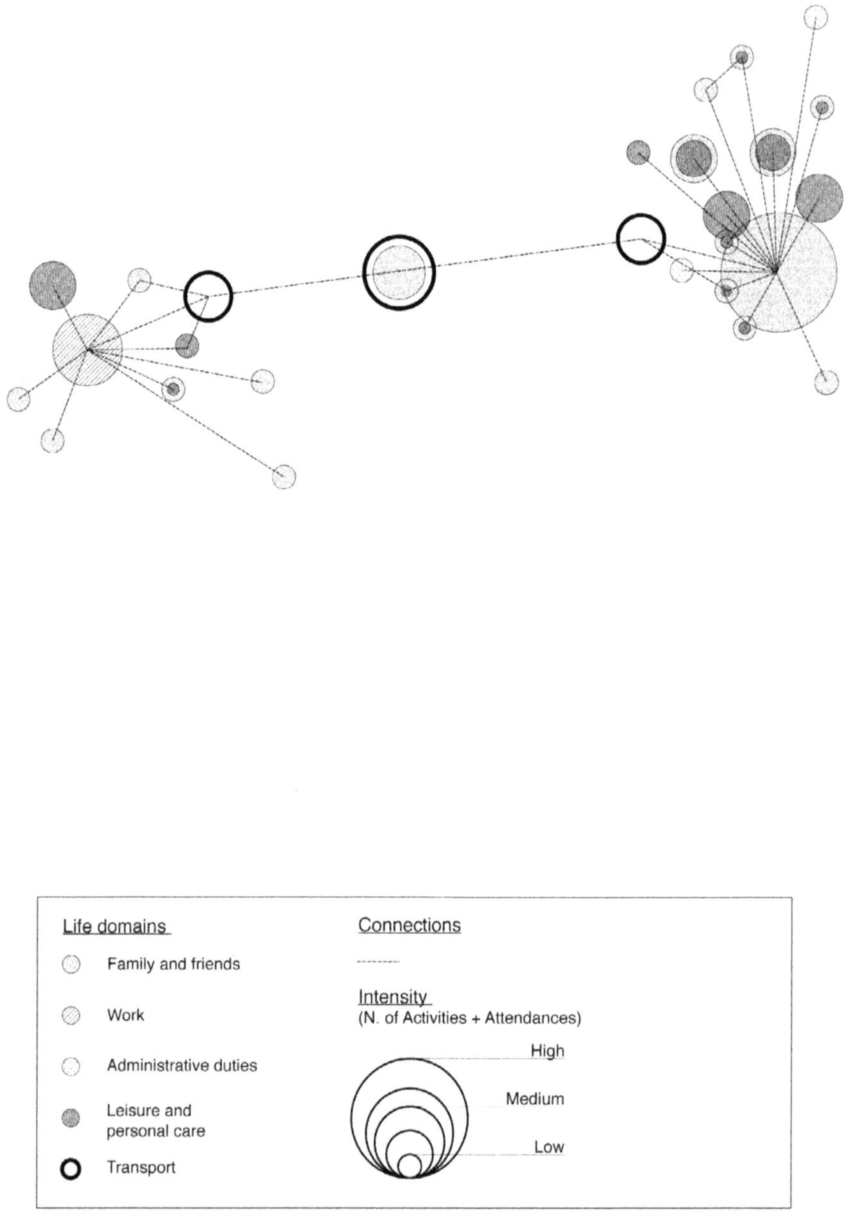


*Figure 4.5* Alessia's map of activities space

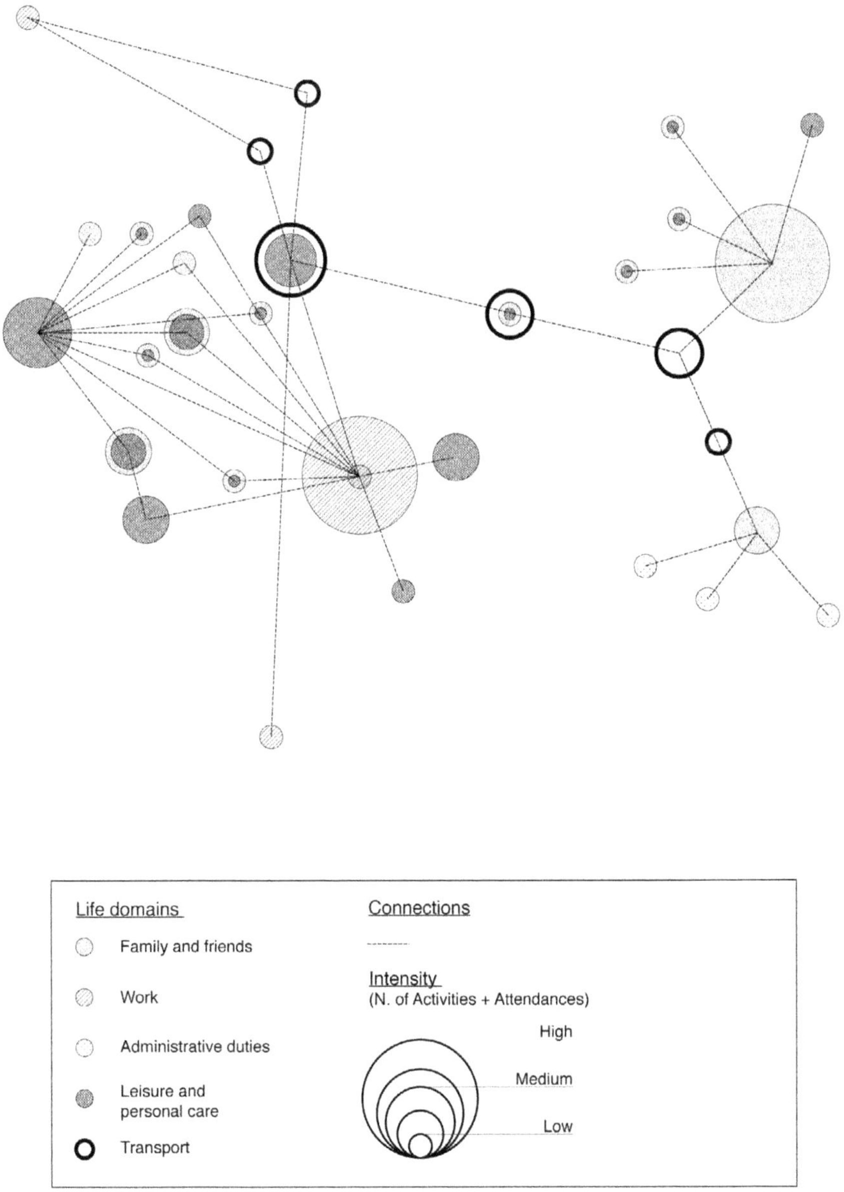


*Figure 4.6* Beatrice's map of activities space

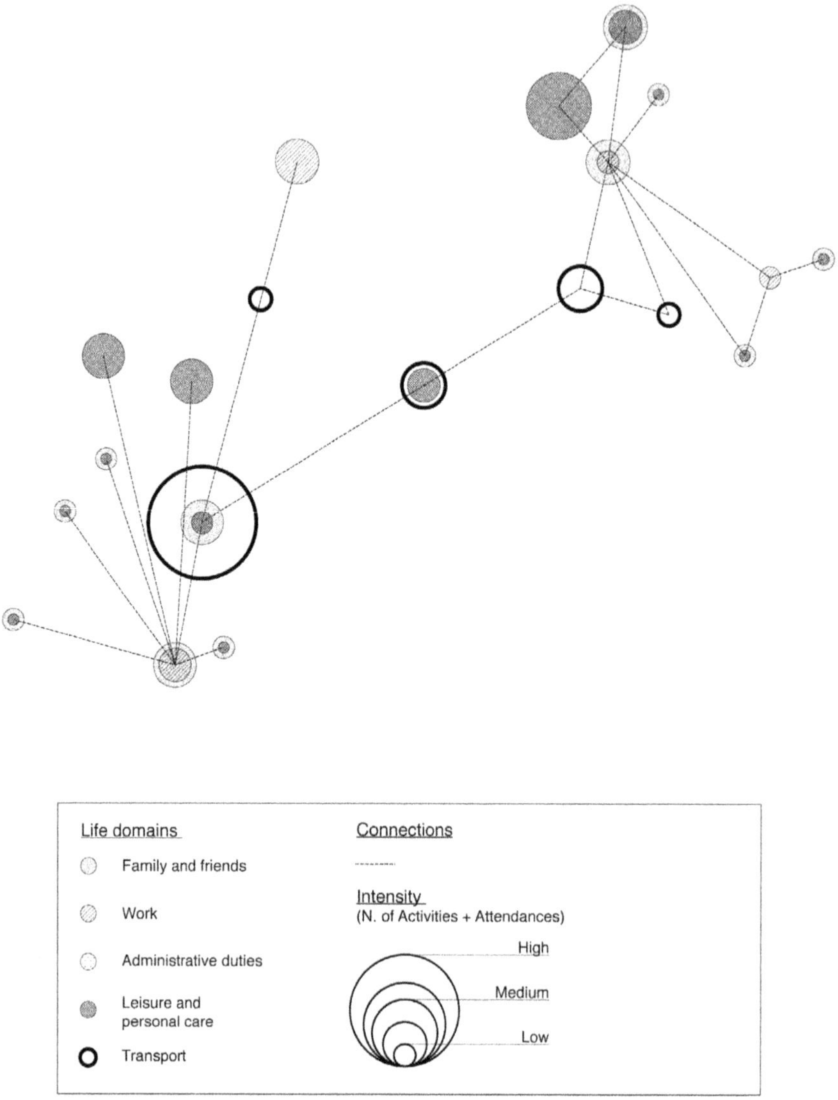


*Figure 4.7* Sandra's map of activities space

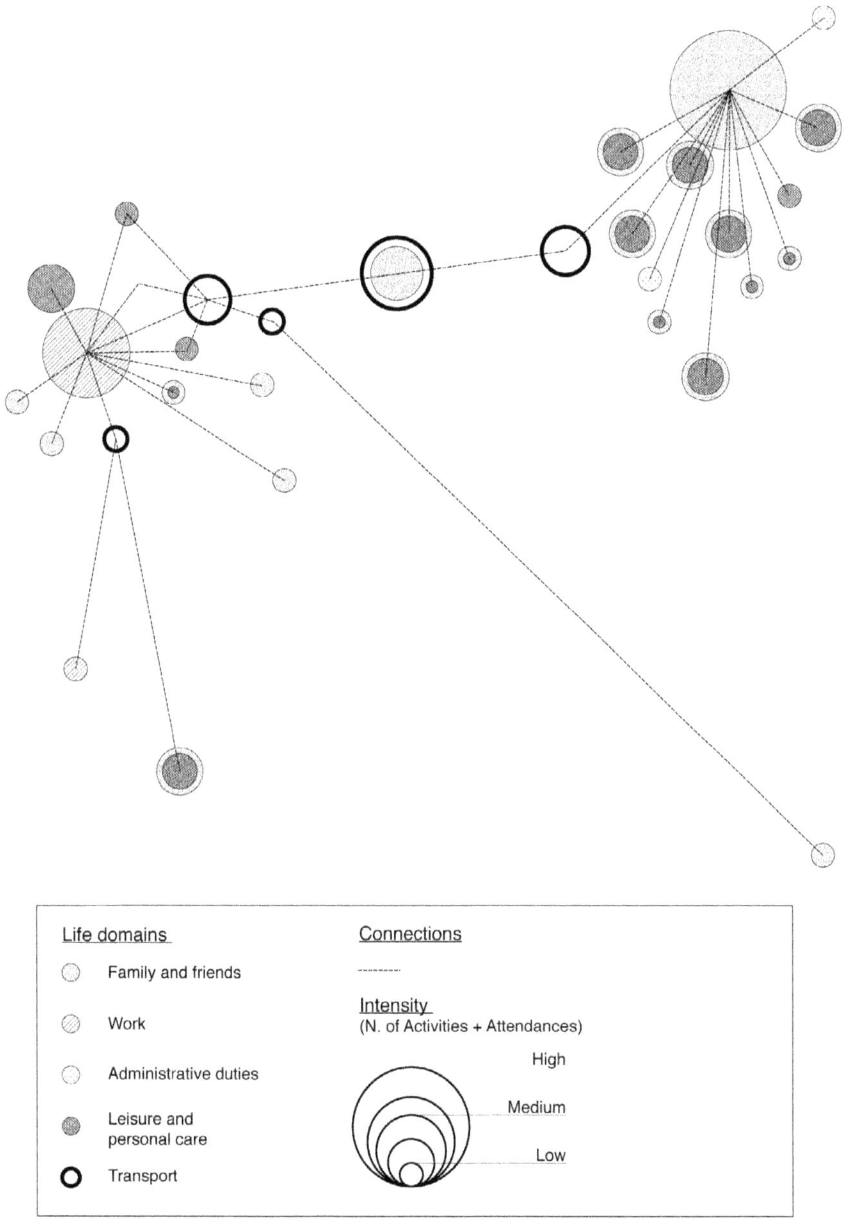


*Figure 4.8* Gennaro's map of activities space

### *Beatrice and Alessia – mobility between job opportunities and personal relations*

Alessia is an LDC between Milan and Turin. Beatrice also used to commute with her every day before she decided to rent a small apartment in Milan and became a shuttle.

In the cases of Alessia and Beatrice it was the specificity of their jobs that led to them being commuters. Alessia works in the sector of public relations for design companies, and Milan is a very attractive city for this because of the many events that are hosted here throughout the year, in particular during the Milan design week. She began working in Milan just as the financial crisis was reducing the number of job opportunities and thus she felt that it was not possible for her to refuse this chance. Subsequently, she has become immersed in her work and being in Milan allows her to follow many important events. Beatrice works in fashion, the company she works for is based in Veneto though Milan is the place where her clients come to see the merchandise and thus it is in Milan that she needs to be.

For Alessia and Beatrice being mobile is a necessity as they will not find an equivalent job in the same sector in Turin, the city where they live. At the same time they prefer to be mobile, rather than permanently moving to Milan, as their significant others, family and friends are in Turin.

### *Gennaro – between high mobility and strong moorings.*

Gennaro has moved for study or work many times during his life. He was born in Fossacesia, in Abruzzo, but he studied in Venice. In 2010 he moved to Turin for personal reasons. His partner was working in Turin and they decided to settle there. Gennaro has been working as a press agent in Milan since 2009 and he does not have much chance of finding an equivalent job in Turin. Gennaro decided to be an LDC. Although he is highly mobile because of his work, and travels often because of work reasons, he is very attached to his house and tries to go back home every night, avoiding overnights for work reasons, because:

> I can't even imagine to move houses and go to live somewhere else. One of the reason why I don't stay often in Milan is that staying in Milan in a condition which is anyway occasional, it makes me more tired than coming back to Turin.

Even though he was living in Milan for many years, and he departs from Milan for his work trips, he can clearly separate this two components of his life as 'Turin: is my relationship/the cultural life/ and the social life. Free time' while 'Milan is the work and part of the everyday duties that I do in Milan because of time'.

***Veronica and Lorenzo – long distance relationship***

Veronica moved to Rome for work, though she is a reporter and often travels all around Italy. In addition, her parents live in Naples and she often visits them. Veronica relocated to Rome, but her partner is living in Milan. Thus when she began living in Rome she started a long distance relationship. Lorenzo, her partner, visits her every weekend, travelling between Rome and Milan. His own work relations have intensified in Rome and because he is a researcher, with the exception of the class he teaches in Milan, he can work anywhere. He therefore spends the majority of his time in Rome.

For Veronica, being mobile was a necessity linked to her job. She is a freelance reporter for Italian television, her job schedules are not fixed, and it can also be the case that she needs to travel twice a week. Therefore it is difficult for her to plan her life. Instead her partner has quite a fixed job schedule and for this reason, in order to maintain their relationship, he started to travel between Rome and Milan. They have a house in Milan and another one in Rome and according to their job constraints they decide where to spend the weekend together.

Veronica likes to travel for her job, but would prefer to be less mobile in her personal life. But as far as moving to Rome was a job constraint for Veronica, Lorenzo's mobility is a consequence of their relationship. At the same time, both of them prefer Rome to Milan because of the urban context, and so they consider Veronica's relocation as an opportunity to spend time in a more pleasant environment.

## Transforming territorial relations

By analysing the relationship that the interviewees have established with the different cities that they use and inhabit, it is possible to highlight the configuration of a less sedentary relationship between people and territory.

The subjects analysed in this research have chosen to undertake more intense mobility behaviours, on a daily or weekly basis, in order to keep their familiar relationships unchanged and to avoid a more demanding migration or residential mobility that could also involve their partners and children. That is to say they have chosen more reversible mobility practices (Kaufmann, 2005). In so doing they have transformed territorial relations.

An important tool in better understanding and describing those transformations is represented by mapping the biographies of mobile people. In particular, the maps of activity spaces help to understand the way in which the interviewees inhabit the territory. These maps show the distribution of different activities on a user-based territory, which we can consider as the activity space (Schönfelder and Axhausen, 2003) of each interviewee. The maps have been built by considering the type and frequency of activities related to a place, while also recording activities undertaken while on the move. In some cases the different activities are equally distributed on the map, in other cases it is possible to identify a preferential relationship with two hubs, in others only with one. This one hub is defined as the 'home', the base to return to at night, and thus the maps already suggest different profiles.

Furthermore, the maps highlight the coexistence of different territorial scales, and the possibility to manage space/time relationships engendering topological relationships (Lyster, 2013) between different spaces, and building new urban rhythms.

### *Three profiles of territorial relations*

Mobility biographies and their translation on the map outline three different ways of relating to territory. Profile A defines people who inhabit an extended space, which includes all of the different places encountered in their everyday life; profile B includes people who can clearly separate activities and places recognising two main centres of life; profile C groups all the interviewees who experience one of the places as being the most important – the 'home sweet home'. And this is independent from the number of other places they inhabit in their everyday mobility.

The mobility behaviours of Valentina, Veronica, Giorgio and Lorenzo have been considered as profile A. The four interviewees have different profiles of mobility: Valentina is an LDC travelling between Turin and Milan every day, Veronica relocated for job reasons, Giorgio shuttles between Rome, during the weekend, and Milan, from Monday to Friday, while Lorenzo is in a long distance relationship and shuttles between Rome and Milan. Valentina has a child, Veronica is in a long distance relationship, while Giorgio is single. They have good economic conditions, and all of them consider being mobile as an opportunity to enhance their professional career. As a consequence of their mobile lives, they develop the ability to find what they need in different places. Furthermore, the fact that they are constantly on the move changes their way of interpreting and experiencing places. Their activities are spread over various places, and the four interviewees have at least three main places they

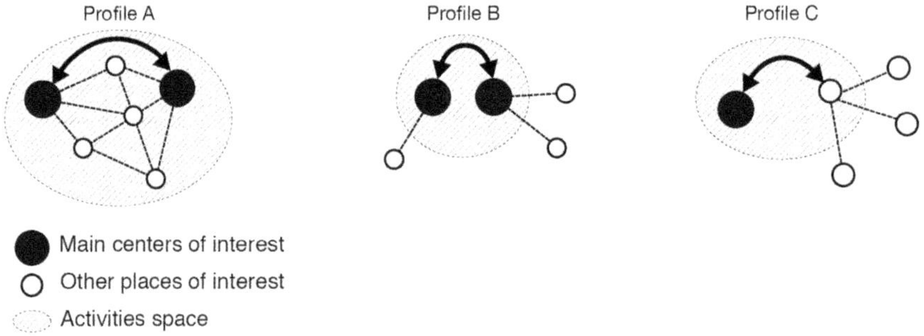


*Figure 4.9* Three profiles of territorial relation

refer to. Veronica, for example, says: 'I am often overlapping the two dimensions; sometimes I think I was doing something in Milan and it was actually in Rome.' It is difficult for her to distinguish between the activities that she does in different places as 'everything overlaps'. She adds: 'I can re-organise my needs according to my mobility.'

Thanks to their economic status they can travel in comfortable conditions. The frequency of their trips is quite different though they all feel at ease while on the move. Thanks to smartphones, laptops and tablets they do many things on the move: they work on the train, they use travel time to maintain personal and professional relationships, they use online banking, shop and send emails. Their aptitude in taking advantage of different places is also applied while on the move. In their cases, 'moving places' become places for 'dwelltime' (Sheller and Urry, 2006: 220).

Beatrice, Alessia and Sandra are included in profile B. They all travel between two different cities. Two of them are in long distance relationships, while one is single. Their friends do not live nearby and they do not have frequent contact. They refer mainly to two places that correspond to two main domains: working and family life. Beatrice, for example, mainly moves between Turin and Milan, although her parents live in Asti and the company she works for has its main office in Veneto. For her, going to Asti is linked to her personal life and so she always departs from Turin, while going to Veneto or abroad for work reasons is part of her working life and so she always departs from Milan. I considered members of profile B as living two parallel lives, and as a matter of fact they all declare this to be the case. For Beatrice, for example, Milan is the city for leisure while Turin is the city for her personal relationship. Furthermore, they do not feel at ease while travelling and try to reduce their travel time.

Gennaro is classified as profile C. Independently of whether he travels between two, three or more cities and if he consider his condition permanent or temporary, he has chosen the place where he wants to live and defined it as 'home'.

He is at ease while travelling but he doesn't consider it to be fun. He is always connected to his partner and significant others thanks to the use of smartphones. He loves to spend time in his main place of residence to go shopping and to meet friends. He prefers to move by bike or on foot while in the city. In the workplace he is not very explorative and all his activities take place close to the office. His satisfaction depends upon the organisation of his trips. He prefers to travel by train because it is faster and because he can relax.

### *Changes in territorial profiles and spatial perception*

The three different profiles are elaborated upon by looking at the activities spaces inhabited every day by each interviewee, and their individual territory.

Profile A people have significantly transformed their relationship with territory as a consequence of their mobility practices. In this case it is not easy to define one place of residence, nor is it possible to say that they are nomadic or unattached to any place. It is more fitting to suggest that they inhabit a larger territory and that they use different places according to different needs. Fast means of transport and mobile devices allow them to adapt to this new lifestyle. The space they inhabit can be considered a rhizome: 'it is a source of opportunities that are in a permanent state of reorganization' (Kaufmann, 2012).

The opposite viewpoint regarding this idea of space is characteristic of profile C. Profile C corresponds to subjects who try to reduce to a minimum the impact of mobility on their personal life. They recognise one place as a unique reference point, no matter how many other places impinge on their life through work, leisure, personal relationships or family. This place is clear and bounded. Nevertheless, changing scale, this space can be linked to other spaces by a network, and so its definition can change from topographic to topological, being defined by its relation with other points of the network (Lévy, 1994: 77).

Profile B represents individuals that tend to identify two places as their main reference, for example home and work. For people corresponding to profile B, travelling between two places means not only returning home or going to work, but to pass from one dimension to another: from work and leisure to family and personal activities, for example.

Definitely the definition of the three different profiles highlighted that the space of high mobile people contains both mobilities and moorings, borders that can clearly differentiate one place from the other as well as networks that can connect all the different spaces.

## The rhythm of the city

Changes in mobility behaviours influenced also the rhythm of the city as different users inhabit spaces according to different schedules made up of presence and absence, on a daily or weekly basis. They have schedules for running errands and shopping, and, in many cases, create the conditions for the immobility of someone else. Those schedules are arranged by the social time of shops, schools and office opening hours, and at the same time are based on the individual and cyclical time of the interviewees. Everyday rhythm results from the interaction of two different temporalities: linear social time and cyclical, chaotic individual time. Developments in transport and the availability of constant internet connections via mobile phones also contribute to changing this relation and to creating rhythms in which a variety of activities can overlap. The choice of different mobility practices has evident impact on the transformation of everyday urban rhythm.

### *Presence and absence: commuting daily or weekly and the rescheduling of traditional activities*

Observing the schedules of the interviewees, it is clear to see that the classic organisation of the week, with services being available five days per week during working hours, does not correspond to the needs of highly mobile people. On one hand, they concentrate personal needs and duties, such as visits to the beautician, gym or post office, during the week in order to have more free time during the weekend to spend with their family or partner; on the other hand, Saturday and Sunday are days full of activities such as sport with children, shopping for the home or cultural activities. Furthermore, when public offices are open, the interviewees find themselves hundreds of kilometres from their place of residence and thus if they need to do some administrative duties, they are obliged to work from home, if possible, or to take a day off. Clearly, although ICT allows the interviewees to stay in touch with family and relatives, physical ubiquity is not possible.

The rhythm of life of the interviewees consists of moments of high concentration activities that interchange with moments of relaxation, presence and absence respectively from the different spheres of activities and places. The rhythm of this presence and absence is determined by the dialogue between the temporality of their travel and their work and family schedules. While shuttles normally use the workplace at night, everyday commuters tend to do many activities during lunch time. While some places and cities are inhabited 24/7 due to the continuous presence of new users, mainly invisible to the lens of official data, other places are experienced in a less intense way. The rhythm of the city is also shaped by the variation of presence and absence, participation in public events and private life, concentration in some places and lack of presence in others.

### *Gender, mobility, immobility and the management of the household.*

The co-presence of mobility and immobility in the household was visible, during the research, in the couples with children where the mobility of one partner is reflected in the immobility or lower mobility of the other.

Presence and absence influences the reorganisation of traditional schedule activities such as shopping for the home or, more generally, the management of the house, causing additional changes in everyday rhythm. While women care more for the organisation of house and shopping (all the women interviewed consider the organisation of shopping as a priority to fit into their complex mobility schedule, and found good strategies), men did not take this concern into account. To this regard Alessia says: 'the shops are all along my way home, but mainly I do shopping during the weekend, and I cook for the whole week. I also have a housemaid that helps me with housework once a week.' Veronica adds: 'I know the schedule of all the businesses in the area so that I can take advantage of dead time to go shopping for the house.'

In this case the weekend is the opportunity to concentrate on various activities: Valentina goes swimming with her son every Saturday morning as this is the only way for them to do sport together, and she also visits museums and exhibitions with him. During the interview Roberto said: 'my son plays tennis on Saturday, in this way I can take him there and spend time together. But it is my wife that takes the kids to school and goes to pick them up every day.'

Definitely it is always possible to observe a dynamic of mobility/immobility in a couple: for every mobile component, in a couple with children, there is always another component which is more immobile and is ready to cover everyday duties. As a matter of fact the presence of one of the partners in a certain place implies an absence in other life dimensions and in another place of his or her dispersed territory. It is therefore the other partner who simultaneously covers these aspects and occupies this space.

The reasons behind this process are quite complex. In the case of Ettore and Roberto, their wives are in charge of the management of house and children. Both are working mothers, but a flexible working schedule and the geographical proximity of work and family make it possible for them to manage family and work. In the case of Valentina, the prevalent cultural and social imagination which considers women more immobile and linked to land and home is inverted. The woman is more devoted to career and travels for work, while the husband has a more flexible job, and takes on the management of the child: 'my husband manages all the questions related to my son,' says Valentina.

Nevertheless, on closer inspection, Valentina and her husband decided to settle in Turin so that their son could grow up close to his grandparents, given that in Milan they have no family support. Furthermore, Valentina has an open-ended contract, while her husband is a building contractor. Consequently, her job is more secure and her salary is higher. These are the same economic and familial conditions that lead to the residential and mobility choices of Ettore, Roberto and Maurizio. These three interviewees have open-ended contracts, while their partners are freelance or run a small business. Therefore, their mobility is a choice that was made due to the economic balance of the couple and because it was also supported by the choice of their female partner to be less mobile.

## Conclusions

In this chapter I have observed the case studies with particular reference to the way in which their mobile lifestyle has consequences on the relationship with territory and on the configuration of urban rhythms.

Presenting the story of the interviewees I have introduced some of the reasons why they have decided to be mobile, making it clear that even though they are a small group of pioneers of a new mobility style, they are not a mobility elite. Rather, in many cases they are obliged to be mobile by their work and by a complex system of personal and family relationships and needs. Their mobility behaviours configure new urban rhythms, made up by the relationship between presence and absence of people and their different time schedules, and by different intensity of use of spaces. These new rhythms also call for new urban schedules.

Furthermore, the biographies of mobile people and their graphical transposition on the maps outline three different territorial profiles: profile A defines people who inhabit an extended space, which includes all of the different places encountered in their everyday life; profile B includes people who can clearly separate activities and places recognising two main centres of life; profile C groups all the interviewees who experience one of the places as being the most important. The different profiles highlight the coexistence of mobility and moorings, boundaries and networks on highly mobile people's everyday life.

These reflections on territorial relation as well as the consideration on urban rhythm are at the basis of the identification of spaces of mobile people and spatial transformation due to emerging mobility practices, as presented in chapter five.

# 5 What spaces for highly mobile people?

## What spaces for highly mobile people?

> On the train I don't work. I read a book and I chat with friends.
>
> Maurizio

In the station I have breakfast with my friends. I can also meet friends for a drink before going back home, and I go shopping with them.

Valentina

I always meet friends on the train. The train is like a lounge.

Giorgio

Even when my friends come to visit me from abroad I meet them at my office.

Beatrice

These are some of the spatial considerations made by the people interviewed in this research. It seems clear that the spaces involved in their everyday mobility are numerous and differentiated, and that even the spaces that have always been considered as mobility spaces, such as the railway station or the train carriage, have assumed more complex meanings.

Mobility spaces or 'mobility environments', to use the definition given by Bertolini and Dijst (2003), are 'places where mobility flows interconnect' and where accessibility is combined with proximity (Bertolini and Dijst, 2003: 28), meaning accessibility not only as a component of transport but also as the availability of activities. Bertolini and Dijst listed airports, railway stations, motorway service areas and pedestrian/bicycle nodes as mobility environments.

The first hypothesis to be discussed in this chapter is that it is possible to extend the concept of mobility spaces to other unexpected places, not linked to flows but responding to the advanced spatial requirements needed by mobile people.

As a matter of fact, this research demonstrated that intense practices of mobility are no longer only the prerogative of rich business men or elite travellers. Instead, due to increasing job insecurity and the need to remain connected for economic and family reasons, there is a growing number of people, in completely different economic conditions, who need to travel and generate new and innovative mobility practices. The rescheduling of urban activities and services, the rise of new spaces for social relations, the need for more flexible spaces, and the change in the way people move toward those spaces, call for new spaces and services that can support them. The analysis shows, for example, that spaces for living and working have shrunk and overlapped, the boundaries between different domains of life are no longer fixed. Furthermore, the disassembling of both temporal and spatial traditional urban structures causes domains of life once considered completely separate such as home and office, or café and office, to overlap.

The new spaces are, for example, the temporary houses needed when shuttling between two cities, or the third places needed to work from home when it is not possible to reach the office, or even the office spaces that are transformed in order to allow a more differentiated use of space.

Looking instead at traditional mobility spaces, I suggest that they are undoubtedly 'generating new opportunities for human interaction' (Bertolini and Dijst, 2003: 28), being appropriate meeting places that contribute to the constitution of network capital (Larsen, Axhausen and Urry, 2006: 280). This is not only because 'flows interconnect' (Bertolini and Dijst, 2003: 31) in those places but rather that, firstly, they are the place where space-time constraints are minimised as they allow an optimisation of time use; secondly, they are polyfunctional and can host many different activities; lastly, they are also considered as landmarks for mobile people, places where mobile people can feel at home while moving between different 'homes'. Furthermore, as observed by Nadler the regularity of travels transforms 'routes in routines' (Nadler, 2014: 273), thus spaces of mobility become familiar places for mobile people; they assume more complex meanings and are not considered simply a means of transport.

In the following paragraphs, I analyse the different spaces and their transformation due to the impact of emerging practices of mobility distinguishing between traditional mobility spaces and common spaces that support the development of mobile lives.

## The railway station: preliminary remarks

Most of the interviewees involved in this research travel by train, and, during the interview, they highlighted the importance of the railway station in their everyday lives. In the following paragraphs, considering the different biographies and experiences of mobility, I will underline how the use of the station has changed due to the presence of new mobility practices, which spatial characteristics are needed for these new uses and which are already available.

I must clarify that the stations that I personally visited while travelling along with the interviewees, and the ones that they mentioned, are essentially the stations along the high speed rail network. These are the stations of Torino Porta Nuova, Milano Centrale, Roma Termini and Napoli Centrale.

The considerations exposed in this research will focus on Trenitalia services as none of my interviewees was travelling with Italo.

### *The railway station as an appropriate urban meeting place*

> My activities are all in the station: Milano Centrale or Garibaldi. I have breakfast with my friends here, as many of them live in Milan. For example, this morning I had an appointment with a friend who works as a photographer, in the end I didn't manage to do everything … so I decided to meet him next week. Normally, we meet in Garibaldi and we have breakfast together. I also do shopping; I buy presents for my friends … Tonight I'm going to meet a friend. We met in a French class a long time ago and since then we've been very close friends. Then she moved to live in Bergamo. We meet at the central station as she comes from Bergamo and I am travelling to Turin. Tonight I'll go to the central station, I'll meet her, we'll have a quick drink together and then I'll go back to Turin.
>
> Valentina

Many of the interviewees consider the railway station as an important place in their everyday mobility. Many works have thoroughly illustrated the particular roles performed by railway stations over different periods of time: from being the gateway to the city (Pucci, 1996: 62) at the beginning of their history, to intermodal hub and opportunity for urban regeneration (Pucci, 1996: 66) more recently. According to the interviewees, it seems that nowadays railway stations have primarily become a preferred urban place to enhance human interactions, where it is possible to meet friends and colleagues for leisure and work, as well as being a place for unexpected encounters, or even a place to have a break between other engagements. This hypothesis is also supported by the European Charter for the Development of Social and Societal Initiatives in Train Stations, which that defines railway stations as a 'central point of urban life … available to all clients and local residents' (European Charter, 2010).

One of the main reasons for the rising importance of railway stations in the urban context is due to the fact that by using railway stations as a meeting place it is possible to minimise time constraints. Valentina, for example, says: 'I don't care about the place. I love spending time with my friends. So the station can be a nice place to meet people.' Valentina declares that what she values more is time, she doesn't care about spatial characteristics as long as she has more time to spend with significant others. Meetings in the station can be squeezed in between other appointments and activities such as transport, work, sport or post office. It can also help in meeting up with people who live in different cities; this is because the station is a connection point between different cities as well as between the city and its surroundings, easily accessible by public transport. It therefore acquires importance as a condenser of different trajectories: the materialisations of 'dynamic simultaneity' (Massey, 1994: 3) that is, according to Massey, the substance of space itself.

A first important consideration of this process is that railway stations cross the border of mobility spaces and become, to all intents and purposes, an urban place and not only a place for transit but a place to rest, to meet friends, to work or to shop. This potentiality of railway stations as appropriate urban meeting places, novel social platforms that promote innovation and exchange (Gwiazdzinshi, 2013: 215), should be appreciated and enhanced when designing stations and cities.

If the aptitude of train stations to optimise the space/time relationship and to maximise space for activities is mainly due to their position in the urban system and their spatial accessibility, there are also other spatial qualities that need to be guaranteed when designing those spaces in order to enhance these aptitudes.

For example, travelling along with Valentina, it was possible to notice that she had carefully chosen the place where we would have a coffee together inside the station. The factors that influenced this choice were different: firstly, the good quality of food and coffee; then the fact that it was clean, and that there was access to a toilet, a comfortable seat, wi-fi and electric plugs.

To this regard Beatrice, for example, says: 'I take a rest in the station, here I know everything. I know where all the different facilities are.' Thus, in order to be a central urban point and an appropriate meeting place, railway stations must be clean, comfortable and secure. Furthermore, communication signs need to be clear, and the facilities well located and signposted.

Many of the interviewees, as regular customers, are able to locate such spatial characteristics inside the railway station. However, when they mentioned the places in which they feel comfortable and have everything they need, it was noticeable that those places have no free access; instead, they are accessible either through payment services – bars and restaurants – for example, or owning a club card. Even though the main Italian and European rail companies signed the 2010 European Charter for the Development of Social and Societal Initiatives in Train Stations, it may be said that amenities such as open access for all, comfort of spaces, free and fast wi-fi connections, access to electric plugs and comfortable chairs do not appear to be valued while designing the layout of stations.

### *The design of the railway station: spaces for transit and the privatisation of spaces for dwelltime.*

> I love spending time in the Freccia Club, I can plan my next trips and book my tickets here.
>
> Giorgio

> It's much better now that I can access the Freccia Club.
>
> Beatrice

Beatrice and Giorgio's perception of the time spent in the railway station has improved since they have had access to the Freccia Club.

As matter of fact, in the renovated HSR stations it is possible to recognise two main types of spaces. On one hand, there are the spaces of transit consisting of corridors, escalators, ramps, where everything suggests the idea of moving forward. There is also the ticket office, the commercial shopping area, the platforms or the subway. Here there are few seating areas in which to wait; they are mainly far from plug sockets, uncomfortable and unsuitable for long waits or to chat with friends or strangers; as a parallel process free access waiting rooms are gradually disappearing. On the other hand, there are the commercial and company lounges which are designed in order to offer comfort, fast wi-fi access, electric plugs, appropriate chairs for working and relaxing, and, more generally, meeting places.

'Freccia clubs' are the business lounges accessible to clients of FrecciaRossa trains. Access is reserved to executive travellers. The train company describes these lounges as the ideal places to relax, work, read and be connected. It is also possible to have small work meetings. In the Freccia Club, seats are comfortable, room temperature is controlled, wi-fi connection is fast and there are plugs sockets available. But Freccia Club lounges are allowed only to elite travellers. Currently there is a paradox: railway stations continue to become more public, appropriate urban meeting points for both travellers and local citizens, while their design and management is oriented to define two spaces: one, open to everybody but transit oriented, poorly furnished and

*Figure 5.1* Milan Central Station: the main hall

*Figure 5.2* Milan Central Station: the gate to the tracks

*Figure 5.3* Milan Central Station: inaccessible lounge

*Figure 5.4* Naples Central Station: Freccia lounge

totally inadequate as a public space; the other, private, access-controlled spaces that are furnished with all the characteristics requested by users, definitely, cutting operating costs and maximising economic benefits.

Looking at the layout of the railway station, it is a relatively easy task to underline all the different boundaries that encircle more or less privatised spaces.

Thus, in the relatively small space of a railway station, it will be possible to see and contextualise what has been defined by Edward Soja as spatial (in)justice (Soja, 2010): only those that have access to private, access-controlled spaces can obtain the 'social resources and advantages' (Soja, 2010: 32) provided in the urban context, while the others have to find their own solutions.

Although the question of spatial justice is not the subject of this research, it is interesting to note that by using a critical spatial perspective on the analysis of the railway stations it has been possible to 'open a fresh look at the subject of public versus private space' (Soja, 2010: 46). Furthermore, I believe that a big design challenge will be that of proposing possible solutions to the process of privatisation being conducted by train companies, and to the consequent privatisation of the space of the station, by designing open spaces that correspond to the needs of railway station users, while at the same time offering an inclusive, safe and comfortable environment.

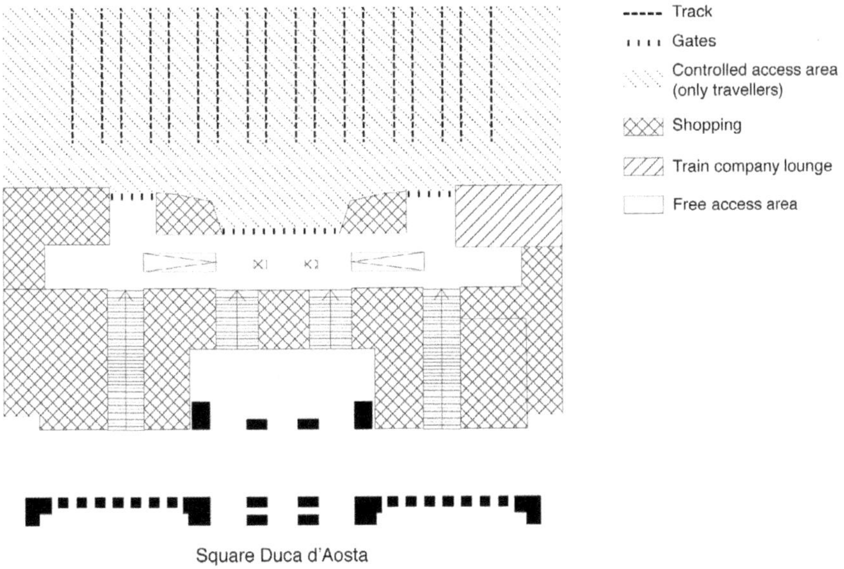


*Figure 5.5* Free-access and inaccessible spaces in Milan Central Station

***The railway station as landmark for urban system***

> The station is a kind of security. The station and the office are reference points for me.
>
> Beatrice

> The station is my place.
>
> Veronica

What is most important to Beatrice is feeling at ease and in Milan central station she is comforted by the idea that she knows the space, and she has some reference points that she can make use of. For her the station is a landmark because from the station she knows how she can go back and forth. It is a safe place while inhabiting different cities, it becomes 'home' while travelling in between different 'homes'.

If Italian cities progressively resemble the horizontal metropolis proposed by Bernardo Secchi and Paola Viganò, which 'has uncertain and evolving limits, independent of administrative boundaries … a typologically differentiated and well-provisioned urban condition' (Viganò, 2013: 97), where different systems overlap and the relationship between individual and territory evolves into a less sedentary way of inhabiting space (Stock, 2013; 794), we may underline that the stations assume the role of social and spatial pivots in the configuration of this new urban condition. They are condensers of multiple trajectories, and landmarks for the constitution of renewed territorial identities.

*Figure 5.6* Roma Termini, the main entrance on Cinquecento square

Unlike the past, railway stations are no longer considered as gateways to the city, when there was an 'inside' the city and an 'outside' the city. For new mobile people, railway stations are more similar to anchor bolts where people take a break while on the move, and they represent an important part in the construction of new territorial identities.

What I define here as a territorial landmark is not a prominent or well-known object in or featuring a particular landscape but rather a 'reference point' in people's everyday lives, a place where they can feel at ease.

It is clearly possible to underline two main ways of interpreting railway stations as landmarks: firstly, and looking at spatial experience, railway stations need to be designed in such a way that users feel at ease and can easily find what they need when they are inside the building. In this way, the station can be a reference point in their mobile lives, a place where they can rest and relax; later, these characteristics can be extended also to the stations' surroundings, to such an extent that activities, shopping and personal duties can be organised in order to accommodate these new urban rhythms.

Rather than simply being monumental buildings, recognisable and visible from far away, stations become an important element in the construction of new territorial identities and a sense of belonging. An example of this change of meaning of stations can be recognised in the competition for the design of the new stations for the Grand Paris subway system, which considered the station as an instrument to enhance a new urban identity. To reach this goal, the winning project[1] suggests an approach based on feelings and users' experience.

### *A public space on wheel: the train carriage*

> I live on the high speed train between two stations. A liquid space between my home and the office, all along the train-line.
>
> Gennaro

> When I am travelling to join my partner, normally I'm really tired, time goes by very fast. I sleep most of the time and I have my relax time. But when I travel for work, the train becomes my office: I work on my laptop, I read the newspapers and I use the internet for research.
>
> Veronica

> When I arrive at the station I meet my train-mates.
>
> Alessia

> I take the train and I start working.
>
> Valentina

Working, meeting friends, reading, listening to music or relaxing. The train can host all of these different activities. In most cases, the train serves as an office to prepare the next meeting or the working day. The train works well for this purpose as people can concentrate on work, but also take a break from personal duties, make a phone call or send an email. In this sense we can assimilate the train carriage to third places (Oldenburg, 1989). In other cases, it can be seen as a place to take a rest or a nap during a very busy working schedule, or as a space for relational activities.

As has already been stated, the people interviewed in this research are frequent travellers who do not correspond to an elite business profile. Most of them possess a monthly ticket and sometimes they travel without a seat, especially during rush hours. They travel mainly in standard class, where seats are 45cm wide and tables, accommodating four people, are 55.5 cm wide for each passenger; nevertheless the space seems to be comfortable enough. On the train 'it is possible to do many activities' says Veronica, while Roberto says 'on the train I can work'. This is an important feature to consider in the design of the train itself. As a matter of fact, the train is considered as a 'place on the move' and not merely as means of transport. This change of perspective, the same that has been applied, for example, to the use of the car, which has been considered as a mobile office or home (Laurier, 2004), can provide some important suggestions for its design.

When I travelled along with Valentina, I met her on the train. Her computer was open on the table and she was doing some work. Valentina explained to me that she had not been able to work at home the evening before, and so she had to catch up before arriving at the office. When I travelled along with Gennaro and Alessia, even though they were not working, I had the chance to meet Cecilia. Cecilia is Giorgio's and Alessia's trainmate;

*Figure 5.7* The train carriage as working space

they usually travel together. Cecilia was working, and concentrating hard to the point that she did not realise who was sitting in front of her. When I asked her if she considered the train as an office she replied without hesitation, 'Yes, sure!'

As I have already stated, the passengers analysed in this research travel mainly in standard class. Nevertheless, standard class is often used as an office. There are some reasons at the basis of this innovative use: the large diffusion and the development of information technologies have completely changed the way of working and understanding work places. This change is influencing not only young freelancers without a fixed office, but also employees. In fact, an ever growing number of companies are supplying their employees with a laptop and/or a smartphone so that their work schedule can be more flexible and employees can work more. Laptops can easily fit into a space of 45cm. As a consequence many employees, as for example Valentina, Cecilia and Roberto, can work on the train when going to, or coming back from, the office. All of the interviewees would like this time spent working in the train to be recognised as part of their work schedule, so that their travel time could be included in their working hours, pushing the idea of smart working.

Travelling alongside with Gennaro and Alessia, it was also possible to see that the train carriage became a space that enhanced social relations. Upon his arrival at the station, Gennaro met Alessia and Carolina – they travel together every day. Sharing this experience has made them all close friends.

Even considering the difficulties of travelling every morning without an assigned seat, they always try to sit together so that they can spend their travel time chatting. In this case, the FrecciaRossa table is akin to a cafeteria, and friends can meet and chat here like in any other public space.

Giorgio, for example, considers the train as a lounge. He affirms: 'I always meet friends and colleagues on the train. Today one of my colleagues is travelling on this same train. If I wasn't travelling with you, I'd have spent my time with him for sure.' What makes the train carriage a perfect meeting place is, without any doubt, the high number of travellers, the fact that it is a 'public space', and the chance to spend time with other people. The four facing seats configuration is a good device to transform the train into a place to chat with other people and exchange ideas. Nonetheless, some people underlined the lack of a proper meeting place on the train. Giorgio, for example, says 'it would be nice if the train company could transform the restaurant into a meeting point. I could go there, I could have a tea or a coffee, and at the same time I could meet people'.

The importance of having a common space in the carriage layout, a place to go for a coffee while travelling, is confirmed by the protests that followed the proposal, made in 2012 by Trenitalia and immediately retreated, to deny standard class travellers access to the restaurant carriage. This reorganisation of the restaurant highlights the intent to have fewer people moving around the train, favouring the delivery of services, rather than strengthening the potentiality of the train to be a real public space 'on the move' in a less sedentary territorial configuration.

### *The ultimate hybrid space: the train carriage.*

> There are a few things that I would love to have on a train: first of all, I love to sleep. Sometimes I arrive on the train and I am very tired. I would like to have a sleeping coach. You arrive on the train and you book your 3 hours of sleep, you get to the end of the trip and you are completely relaxed.
>
> Giorgio

> I am very tired, time goes by very fast. I sleep most of the time.
>
> Veronica

> For me the train is a moment to relax.
>
> Sandra

Additionally, to work, many of the interviewees are tired because of their mobile lifestyle and use travel time to take a rest. As a matter of fact, people affirm, in many cases, that time spent on the train is time for personal care and relaxation. To this regard, Giorgio made a provocative but interesting proposal during the interview that highlights one of the great potentials of the train: he proposes a sleeping carriage. As on the night train, where travellers can make a reservation, even after catching the train, your couchette could be booked for the entire trip or only for a few hours. Giorgio is a shuttle and travels mainly between Milan and Rome twice a week. His proposal might only work for long routes and, of course, would probably be difficult to realise mainly due to the high cost of the service, the complexity of management, booking and cleaning operations. Nevertheless, this proposal highlights a key point: considering that the mobile people analysed in this research are often tired due to their lifestyle, travel time is an important temporal supply for taking a rest.

With regard to everyday LDC, for example, their travel conditions are quite different. As mentioned previously, they generally possess a monthly ticket and have no assigned seat. Beatrice, for example, was a long distance commuter from Turin to Milan; she said:

> the moment I gave up everyday commuting was when they [the train company] were reorganising the train schedule. There were very few trains, and coming back home every day with the 6.05 p.m. or 6.45 p.m. train was impossible. Every morning you had to fight to get on the train, sometimes having to sit on a step for the entire trip. Now I am very happy, I can leave the house at 9 a.m. and get to the office on time. Every morning I see the alarm and I think how lucky I am: at this moment I'd be in Rho, and I'm still in my bed!

Beatrice's experience highlights many important questions: firstly, Beatrice declares that when she changed from being an everyday LDC to a shuttle her fatigue was reduced; nevertheless, Beatrice still affirms that she is often tired. We can therefore say that whether travelling every day or a few times per week, a hyper mobile lifestyle is tiring. Furthermore, what makes mobile people tired is not the discomfort or design of the train but their daily and weekly schedule. Thus, in order to improve the comfort of the train and enhance its potential of being a place for relaxation and rest, it would be necessary to work on train schedules. It may also prove useful to work on time policies[2] in order to accommodate workers' schedules, or improve practices such as smart working to reduce the number of travellers during peak hours.

## A rising demand for more hybrid urban typologies and services

According to Di Marino and Lapintie we are living in a post-functionalist city, 'a city where the boundaries between urban functions have become blurred, where different functions coexist in the same space, and where new and unprecedented functions emerge through citizens' appropriation of places' (Di Marino and Lapintie, 2017: 6). As already considered in this research, laptops, wi-fi, and the development of ICT and transport, have completely changed the classic idea of work space and public space, fragmenting time and places and making everything more fluid and overlapped.

Nevertheless, projects on mobility tend to focus only on public transport policies and transport spaces during the daytime and seldom do they cross the threshold of more private spaces and time. However, if changes in mobility produce transformation in space/time relations and influence urban rhythms, spatial consequences of new mobility practices must also be found in more private spaces and times, as they also contribute to the configuration of those rhythms. After all, Amin and Thrift underlined that the analysis of urban rhythms 'can highlight neglected temporalities' (Amin and Thrift, 2002: 17) and, I will add, neglect spaces.

Some of the interviewees are employees and have a stable office; others are freelance and make reference to at least three or four different workplaces, even in different cities. Other interviewees work from home or from other places; finally some of them have totally unstable job conditions and 'chase' their job, as in Sandra's case. I have already considered how these people appropriate travel spaces and use new services in stations and on trains. With regard to housing and offices, however, these different job conditions, together with diverse mobility behaviours, also generate very different relationships with space and places: in some cases the office is akin to a second home, whereas shuttles need temporary houses that they only occupy from Monday to Thursday or Friday. On the other hand, long distance commuters often work from home at night or for a few days per week, and thus they need comfortable temporary working spaces. This may be at home or in third places, places that are neither the home nor the office, such as cafeterias. Such places have recently emerged as important locations in contemporary city transformation.

### *Working places: redesigning the office layout*

> When I started this job, there were only four of us. The office was an apartment and the atmosphere was very cosy. Then we were bought by the Della Valle group,[3] and now we are part of a French company and we moved to another office. Before there wasn't even an alarm, and I felt as if I could go to the office at any time. Here everything is more official, although I have the keys and I can come whenever I want. I keep my tea and biscuits in the kitchen, I bought them in Turin and I have the same ones at home.
>
> Beatrice

For Beatrice, the office is a second home. It is not only her place of work, but a more complex space where she can spend most of her time, feeling comfortable and at ease. As a matter of fact, when her office became part of a bigger, more institutional and representative organisation, Beatrice was disappointed. This is due to different factors: until two years ago Beatrice did not live in Milan and the office was her second home. Even now that she rents an apartment in Milan, she feels more at home when at the office than when in her apartment. She keeps her things in the office: her gym bag, or tea and biscuits for breakfast; it is in the office that she also meets her foreign friends when they arrive in Milan. Furthermore, Beatrice works in a creative sector (fashion). For her the office must have a multipurpose role: a space where she can work and at the same time relax, or have meetings with her colleagues. Beatrice's perfect office is cosy, and made up of different environments: workspace, kitchen, meeting rooms, places for informal meetings and even places to relax or have a break.

A flexible working environment is a common need and new office layouts are being designed with this trend in mind. The people observed during this research are, once again, a preferred point of observation. For example, for Beatrice, the office has become a kind of second home. Most of the shuttles use the office as a departing point for their trips back and forth, and LDCs spend all their working days inside the office.

In attempting to respond to the request for more flexible working spaces, many companies worldwide have begun to modify the office layout, creating workplaces that respond better to their employees' needs. Firms operating in the creative sector such as Google, for example, have always been attentive to such demands. Nevertheless, the need for more flexible working spaces cannot be tagged only to certain working sectors.

### *Working places: the effect of smart working on urban environment and quality of life*

> An interesting matter is working from home, recently there has been a lot of attention on that, many foreign companies, but also some Italian ones, are developing it. But this opportunity needs to match the role that the employee has in the company. Tetrapak, is a good example, as they allow 1 or 2 days per week of working from home, and there is evidence that people are more productive when they work from home.
>
> Stefano Casati[4]

According to Casati, smart working practices already seem to be a popular phenomenon although they are not institutionalised, apart from certain cases such as Tetrapak.[5] The diffusion of these practices can have many positive effects on cities: such as the possibility of reducing traffic, having a good impact on transport systems and environments. Mobile workers, as reported in the interviews, are advocates of such practices and can be a key population for their diffusion and development.

Smart working can have a large impact on cities and has a vast potential to influence liveability and people's quality of life. The report of smart working days in Milan offers a great deal of evidence that can support this hypothesis.[6] The municipality of Milan launched the first *Giornata del lavoro agile* (smart working day) on 6 February 2014. This day involved 5,681 employees, of whom the majority were LDC. It seems that the most important benefit for workers relates to the possibility of organising their time more efficiently and managing the relationship between work and family. Many considerations were also made regarding the comfort of the workplace, considering their home, or the place they chose to work from on this day, as being more comfortable than the office.

The diffusion and institutionalisation of smart working can certainly have a positive impact on quality of life, it can improve transport systems by reducing the number of travellers during peak hours, it can have an impact on the office space surface by reducing the real estate footprint of many companies, although there will be a rising need for alternative working space as people work from home. Valentina and Lorenzo usually work from home and their need for a temporary workplace is solved inside the house. According to Granata and Lanzani, this overlapping of home and working spaces dates back to the historical concept of the house/workshop, but has led to the idea of the house that can also host a temporary office, especially for freelancers (Lanzani and Granata, 2011). In both cases, this suggests an opportunity, and a need, to redesign office and houses.

### *Housing: diffused micro-transformation of urban environment*

Veronica has a secondary house in the city of her workplace that often becomes an office also for Lorenzo, her partner. Beatrice, Roberto, Ettore and Giorgio need a second house in the city of their workplace. In the latter cases, the house is a part-time house, inhabited from Monday to Friday. None of them have any particular attachment to these places; they simply need to be comfortable, and well connected to the city centre and office. Furthermore, travelling alongside Giorgio I had the chance to meet one of his friends. She works twice a month in Milan, where she rents a house that she shares with a friend. The building's janitor also has a copy of the apartment keys so that he can help them manage the house. Sometimes they lend the house to friends as the apartment is not used most of the time.

Mobile people interviewed for this research normally rent a house for the full week, even though they use it only from Monday to Friday; in some cases, renting a house close to the office is the only way for them to maintain job and family when the distance between home and work, and the economic commitment, does not permit long distance everyday commuting.

In the last decade, in the UK, the trend of Monday-to-Friday rentals has grown incredibly, as witnessed by the continuous appearance of new dedicated internet sites.[7] The rise of this phenomenon is also confirmed by the number of ads on newspaper that rose from 6 to 100 in six months in 2007, together with the age dynamic of the people living in shared accommodation. Apparently, also in Italy the number of people who decide to rent a spare room in their house is growing as witnessed by research elaborated by the University of Siena in 2017.[8] This data suggests the rise of a new mobile population made up of workers that inhabit two cities: one from Monday to Friday and the second during the weekend, and create, at the same time, new urban behaviours and the need for new temporary accommodation.

Those small housing tactics, even though fragmented, do generate micro-transformations of urban environment (Boeri, 2011). And even the space of the house must change, becoming more flexible and variable, in order to constantly adapt to different configurations of inhabitants

### *Services for highly mobile people: Lorenzo, Veronica and 'Enjoy'*

Veronica and Lorenzo are a couple in a long distance relationship. Lorenzo lives and works in Milan while Veronica works in Rome where she rents a house. When Veronica temporarily moved to Rome, they decided to rent an apartment that would also be comfortable enough for Lorenzo to visit. Lorenzo owns a car in Milan, Veronica another in Rome. When they want to meet each other they travel by HSR, and when they arrive at the station they take an Enjoy car and reach the other's house.

Enjoy is a car sharing services created by Eni, together with Trenitalia, Fiat and CartaSi. It started in 2013 in Milan and since 2014 has also been available in Rome. This is one of the reasons that Lorenzo and Veronica like to use it. Even though they have a car in both cities, 'Enjoy' allows them to be completely independent. In addition, they do not have to pay congestion charges in either Rome or Milan, they can park for free, they don't need to return the car to a specific point and can leave it anywhere in the city. Furthermore, the subscription is free of charge and is valid nationally, so that with the same code Lorenzo and Veronica can rent a car in either Milan or Rome, while paying only one subscription (which is not the case for public transport passes or bike sharing services).

There are certainly many criticisms of this type of service, however, mostly linked to its implementation and management. Regardless, similar services, such as car2go which is already active in many European cities, are currently being implemented. It seems that new generation car sharing, such as Enjoy, has touched upon some important issues: having one registration code that enables people to rent a car in different cities, a common system of registration and payment, free parking even in the city centre and free access to congestion charge areas.

## Conclusions

Thanks to the analysis of mobility behaviours, it has been possible in this chapter to highlight some important spatial changes. Describing the individual paths of a group of new commuters, and mapping their lives and mobility, has provided a greater insight into the changing of urban rhythms generated by their way of inhabiting space as well as highlighting their continuous re-negotiation of boundaries between different territories.

Even though case studies are limited, it is apt to consider that the transformations engendered by mobility behaviours on urban space may be quite important: 1) their influence is extended to a wide variety of spaces and uses of the city, including the need for temporary houses, mobile office/spaces, or new uses of public spaces and transport; 2) their use of space influence a large number of people including their families, relatives and friends; 3) because of their mobile lifestyle, the people observed in this research are 'pioneers' (Kesserling and Vogl, 2008; 5) of innovative urban behaviours and can be considered as niches (Geels, 2012) for the development of new lifestyles.

From one side, the use of an actor-based approach highlighted that traditional mobility spaces acquired new meaning for highly mobile people. The railway station is not only a place for transit but also a place for 'dwelltime' (Sheller and Urry, 2006: 220), which allows the optimisation of space/time relationships and consequently is chosen as a preferred urban meeting place; at the same time, it is a landmark in a territory made of geographically dispersed activities. Time spent on the train carriage, instead, represents a significant time supply that can allow people to meet and perform various activities, and thus the train carriage cannot be considered as a container where immobile people are transported from one point to another. Instead, it is a veritable public space on wheels. Kesselring defines those spaces as 'connectivity spaces' (Kesselring, 2006: 277), places for 'interaction and contact', where 'travelling time is not dead time between origin and destination'; instead it is a time to live the experience of movement.

On the other hand, numerous places, traditionally considered not to be mobile, are transformed by mobile people in order to better respond to their needs while searching for more flexible and hybrid spaces able to support the construction of their mobile lifestyle. Office space shrinks due to smart working, and becomes more flexible, with spaces for work meetings, concentration and relaxation, and third places become ever more common in urban and rural context. At the same time, there is a rising need for secondary houses used only from Monday to Friday. This process influences the real estate market and creates a turnover of high and low intensity activities at the urban level. Services such as smart working and car sharing are helpful tools to adapt to this continuous rhythm of presence and absence.

All of these changes are activated by mobile people in order to enable their mobile lifestyle. Thus, mobile people configure a city made up of the simultaneous presence of very different and extreme intensities of use (Viganò, 2010: 31), where surfaces can shrink, as for example in the case of the office, and houses can be deserted during the weekend. At the same time spaces can be inhabited and services can be used in a very intense way. In this sense, urbanism is considered not only as the discipline of material and permanent city design, but also as the coordination of these multiple urban rhythms made up of the alternation of different uses.

## Notes

1 This information is issued by the report 'Architecture and design des gares du grand Paris. Résultats de la consultation de la mission de conseil' [Architecture and design for Grand Paris Stations. Results of the competition] (Société du Grand Paris, 2012).
2 Urban time policies are public policies that intervene in time schedules and time organisation regulating human relationships at the urban level. They have been implemented by the Italian municipalities according to national law 53/2000 about parental leave and time management in the cities (Bonfiglioli 2001).
3 Della Valle is an Italian firm located in Marche Region, it produces shoes, bags and clothes for different fashion brands: Tod's, f=Fay, Hogan and Roger Vivier: http://www.todsgroup.com/it/group/history
4 Stefano Casati is an Italian entrepreneur, co-founder at DoubleYou srl, now part of Gruppo Zucchetti. Passionate about technology driven ventures, disruptive products, process innovation, DoubleYou is a society specialised in welfare and mobility management.
5 Tetrapack, together with Mars, was awarded the smart working award in 2013 from Osservatorio Smart Working – School of Management del Politecnico di Milano
6 I refer here to the report 'Gli esiti della giornata del lavoro agile. Una sperimentazione del Piano Territoriale degli orari di Milano' [The results of Milan smart working day: an experiment of Milan time working plan] (Comune di Milano, 2014).
7 www.mondaytofriday.com, www.spareroom.co.uk or www.fivenights.com
8 The aerification of cities, a research elaborated by Ladest Lab about the diffusion of Airbnb in Italian cities: http://ladestlab.it/maps/70/airbnbscapes [28 November 2019]

# 6 Conclusions

## Research findings

How have changes in mobility practices transformed and reshaped spaces and their uses in contemporary cities? In the process of answering these questions this research reaches the following outcomes.

First of all, the research enters the theoretical debate regarding emerging mobility practices offering a particular regard to reversible mobility. Reversible mobility practices are those that allow an individual to combine in everyday life places of activities that were before spatially impossible to assemble (Vincent-Geslin and Ortar, 2012: 40); a concept which has been contentiously accepted in the field of mobilities studies.

To better analyse those forms of mobility, a qualitative method has been implemented in response to the need for less static methods of mobility research and a more interdisciplinary approach, based on the direct observation of mobile people. This method allows us to consider that the decision to be mobile is linked to economic needs related, in most cases, to family and relational constraints. This finding is important in order to disregard two common ideas about mobility. Firstly, the idea that mobility is an essential requisite of the contemporary world (Kaufmann, 2011; Cresswell, 2010b) together with a narrative of mobility 'as education, as freedom, as modern' (Cresswell, 2010b), showing instead that in many cases mobility is not the result of a choice but a basic need. Secondly, the idea, congenital with a technological determinism, that considers an increase in mobility as a simple consequence of ICT and transport development, demonstrating that mobility has to do with everyday behaviours. The method also furnishes a sensitive approach to mobility spaces that permits an understanding of the way in which mobile people experience those spaces and considers the rising demands that are generated by mobility practices.

Thanks to this methodology, it has been possible to interpret spaces of mobility not only as those spaces affected by large flows of mobile people but as all the spaces that support the development of emerging mobility practices. It also allows us to observe the changes occurred in the relationships between people and territory and to sketch three profiles of territorial relation.

## From mobile people to spaces of mobility

As considered in this research, catching mobility in its complex network of relations and simultaneity proved difficult when only using transport data. At the same time, analysing mobility through the observation of transport spaces does not allow all the systems of spaces and services required to support a mobile life to be highlighted. This is the reason why I decided to adopt an experimental research method. In defining this method I was inspired, in particular, by other methods such as drive-alongs (Laurier, 2010), being 'mobile-with' (Bissel, 2009, 2018) and performed interviews (Latham, 2003).

Travelling alongside with mobile people I was able to cover hundreds of kilometres in the same day and, at the same time, share the small spaces of the train with them. This facilitated an understanding of different spatial implications regarding changes in mobility practices, and offered an insight into the different spaces involved in an interviewee's everyday life. In this way I could extend the definition of mobility spaces to all those spaces that are used to support a mobile life.

This method allows me to describe reversible mobility practices and consequently a reversible use of spaces. Reversible spaces are spaces that can be subject to interpretations that are completely different and disconnected from one another (Bourdin, 2013: 185). In fact, looking at the analysis conducted in this research, it is clear how some services and places may be used by different people at different times, temporarily co-used and reconfigured from time to time. Consequently, their meanings can be reversed by the different users each time.

In this sense, reversible mobility configures a city made up of the presence of very different and extreme intensities of use (Viganò, 2010), where surfaces can shrink and houses may remain empty at weekends. At the same time, spaces can be inhabited and services can be used in very intense ways. In this sense, urbanism is considered not only the discipline of material and permanent city design, but also as the coordination of these multiple urban rhythms which are made up of the alternation of different uses.

## New territorial relations

Looking at the relation between people and territory, the research elaborated three different profiles. As considered in chapter four the three profiles correspond to different ways to build an individual activity space. People responding to profile A have significantly transformed their relationship with the territory as a consequence of their mobility practices. In this case, it is possible to say that they inhabit a larger territory and that they use different places according to different needs. Fast means of transport and mobile devices allow them to adapt to this new lifestyle. The opposite approach is characteristic of profile C. Profile C corresponds to subjects who try to reduce to a minimum the impact of mobility on their personal life.

They recognise one place as a unique reference point, no matter how many other places impinge on their life through work, leisure, personal relationships or family. This place is clear and bounded. Profile B, instead corresponds to individuals who inhabit a space made up of two main points, tending to identify two places as their main reference, for example home and work. Those profiles may be relevant also from the operational point of view suggesting three different solutions of mobility design that matches from time to time the needs expressed by different profiles, as also suggested by Pucci and Vecchio with regard to Milan Metropolitan Area (Pucci and Vecchio, 2019: 8)

## Open questions for a design agenda on mobility spaces

The analysis of highly mobile people's practices and needs is open to further questions that may be the subject of a research agenda on mobility which looks at people-focused and needs-based (Lucas, 2012) solutions.

An important question is linked to the management of administrative borders and ticket pricing, both having consequences on citizenship rights. Such issue was raised during the interviews at various times and moments. For example, most of the interviewees complained about the elevated cost of using different transport services both within different cities and between them. The diffusion and implementation of integrated and smart ticketing systems nationally could resolve this problem, and initiate coordination and cooperation between different transport companies.

Also the configuration of new territorial relations raises the question of 'who pays for what?' Since Beatrice left her parents' house 12 years ago, for example, she has never changed her administrative residence. She has, however, lived in various houses in various cities. This condition causes her many problems linked to her citizenship rights. They range from more mundane issues regarding her administrative and banking affairs, which are associated with her official residence, to more complex matters such as the right to vote or the right to public health care. Furthermore, it partially excludes her from the decision making process in the other places she lives and works. Clearly, this research is not the place to thoroughly analyse these questions. However, an interesting approach has been offered by Crosta (2003) in this regard, quoting Gerald E. Frug (1999), with respect to a polytopic system of voting which is based on actual belonging rather than on administrative residence. Nevertheless, it has had no operational follow-up and thus the question remains open.

A last question is related to the management of private/public spaces. The term privately owned but publicly used spaces (POPS) has been:

> first used in the 1960s in New York City. As a legal oxymoronic invention, it comprises two parts: 'Privately Owned' refers to the legal status of the land or building in which the public space is located .... Thus, it is a

> 'Public Space' rather than a public property in this case since it is not owned by the city.
>
> (Kayden, 2000)

As matter of fact, according to Mitchell, what makes a space public is not its 'preordained publicness' (Mitchell, 2003: 35), rather it is the use that some groups make of such spaces. Public uses of arcades and commercial shopping malls have been a central subject of discussion with regards to these matters. This research highlighted that the spaces of the railway station, as well as the spaces of the train carriage, can easily be related to this typology. As stated previously, the activities that can take place in the railway station, and mobile people's familiarity with such spaces, definitely transform the station into a public urban meeting place. Nevertheless, they are privately owned and managed. This creates a fracture between 'dwell-time' spaces and spaces for transit. The former are subject to controlled access and designed in order to accommodate customers, while the latter are designed in order to keep users moving. This differentiation is due to security requirements which tend to be particularly tight in station buildings. It is certainly possible to say that the transformation caused by changes in mobility practices in station spaces have rendered such spaces more public, and thus also subject to management problems traditionally linked to POPS, such as control and security.

Despite this, these concerns raise a question about spatial injustice as it creates a city made of different parts where access is dependent on different economic, social and cultural possibilities.

# 7 Post-face opening

## Working on mobility from people

*Voncent Kaufmann*

For the past 15 years or so, many mobility surveys have highlighted that when measuring all travel, the average daily time budget allocated to it increases in most European countries, thus contradicting the 'Zahavi conjecture'. What is behind this upward trend? A new balance between anchoring and mobility? A new attraction for mobility?

Mobility is at the heart of our lives, it helps to define its framework and increasingly it becomes a structuring dimension, as Bruna Vendemmia brilliantly showed in the preceding chapters:

- The daily life of households has become more complex, with on the one hand the multiplication of the spheres of activity of its members (with the dual professional activity, leisure and extracurricular activities of children), and on the other hand, the spatial fragmentation of the spaces of daily life. The result is an increased spatial and temporal tension in activity programmes, of which mobility is a central element: quality of life will often depend on how mobility is programmed and the ingenuity of the selected sequences.
- Flexibility is now a requirement of the work market that often involves mobility. With the increase in fixed-term contracts, flexible working hours (and even the return of on-call work in certain professions), stayings abroad and more generally business travel, the ability to be mobile is essential. This situation is reinforced by high unemployment rates: in such a context, an employee is satisfied with a job even far from home.
- We are witnessing a multiplication of ways of moving through time and space, thus ensuring the co-presence of beings or actors. There are strategic choices and distinctions where there was once only the constraint of spatial friction. In addition, technical and social innovations are constantly changing the access, skills and ownership that make it possible to be mobile. As a result, individuals and social groups must continually adapt. On the one hand, the multiplication of possibilities introduces differentiation where previously there was none. The actor is therefore increasingly confronted with choices of access (to which it is necessary to

acquire or not), skills (to acquire or not) and appropriation (analysis of the interest of a particular means of communication) when he wishes to be mobile.

These developments suggest that the growth in mobility refers to major societal changes. Yet very often, current approaches to mobility still do not provide a satisfactory understanding of the ins and outs of mobility growth. As a result, the knowledge of this growth is impressionistic and incomplete.

Describing, understanding, explaining and regulating mobility requires transversal approaches.

When a geographer talks about mobility, he or she is not talking about the same thing as when an engineer or sociologist use this notion, which makes it difficult to establish a dialogue between their respective knowledge. In other words, when we talk about mobility, we do not know exactly what we are talking about: everything depends on the discipline from which we come. Mobility is social and spatial, physical, virtual or potential and it concerns people, goods and information.

Even by reducing the focus on the spatial mobility of people, four different meanings of the term mobility are used in human sciences to describe movements in geographical space: residential mobility (referring to the residential pathway), migration (referring to international and interregional movements of emigration and immigration), travel (referring to tourism) and daily mobility (referring to movements of daily life).

Traditionally, mobility analysis proceeds by partitioning into four differentiated forms according to two dimensions: the temporality of which it is based (long temporality, short temporality) and the space in which it takes place (internal or external space to the living area). Each of these forms of mobility is the subject of an abundant literature and a specific disciplinary base. If daily mobility has been generously studied by geographers, the analysis of residential mobility has often been the prerogative of demographers, anthropology has appropriated the journey, with migration traditionally remaining a sociological field.

Therefore, each form of spatial mobility refers to a specific field of research dealing with a specific object. As a result, it is not easy to deal with cross-cutting research objects that do not fall within these definitions. On the other hand, each of these forms of mobility is the subject of an abundant literature, but sometimes giving the impression of trampling: research is constantly being carried out on well-known sectoral themes such as the anchoring of daily mobility practices in lifestyles, residential mobility and the life cycle, modal practices, without any real scientific progress being there. However, it is the links between the different forms of mobility and the trade-offs made by the actors that fully reveal the phenomenon of mobility and their stakes.

With the rapid and massive expansion of the spread of information and communication technologies, the practice of each of these mobilities is changing to the point that they are becoming blurred. Distances and speeds are

*Table 7.1* The four classical forms of spatial mobility

| | *Short duration* | *Long duration* |
|---|---|---|
| Internal to a living area | Daily mobility | Residential mobility |
| To the outside of a living area | Travel | Migration |

increasing, while the temporalities to which they refer are desynchronising and re-synchronising themselves in new ways. Thus, new forms of mobility are interspersed between the above four by combining them, as illustrated by the following examples:

- Bi-residentiality (between daily mobility and inter-regional migration or residential mobility). While the seasonal double residence has existed for a long time, its development on a weekly basis is recent. It covers very different situations. Among dual-activity couples, it is often the result of family arbitrations when the spouses' professional activities are not located in the same urban area. There are also dual residence practices with second homes inhabited three days a week. Alternate custody of children after divorce is another example of the developing of dual residence.
- When the workplace is very far from home, another practice other than dual residence is currently developing: the very long distance commuting associated with working from home (between daily mobility, residential mobility and travel). It consists of travelling to work only one or two days a week and working at home the rest of the time. This practice is largely based on the possibilities of working at a distance provided by electronic messaging.
- Short-term tourism (between daily mobility and travel). Leisure mobility between holidays and weekends has grown considerably over the last decade. They often take the form of discovering a city and its cultural richness in a relaxed way.

The common feature of these new forms of mobility is that they are largely based on the use of information and communication technologies. They play on the combination of speeds between transport and communication systems and the ability of stakeholders to appropriate technical systems. They often consist in giving up developing forms of movement and avoiding choosing between alternatives such as moving or not taking a new job, being in a relationship or living alone, etc.

As research on spatial mobility has been structured around the four main forms we have mentioned, it is difficult to identify and even more difficult to analyse these emerging forms. De facto, the differentiation between daily mobility, residential mobility, travel and migration is losing its relevance and needs to be reviewed: the multiplication of travel possibilities and the increase in speeds introduces new possibilities.

As we have just seen, spatial mobility is not really a concept in the social sciences, but a notion that is fragmented by fields of research and disciplines, vague because it covers phenomena of a different nature without any links between them, and this leads research to neglect the transversal forms of mobility, which are precisely those that are at stake.

To address these limitations, we propose to develop a new conceptual tool to capture spatial mobility that focuses on the effects of interaction between different forms of movement. This does not start from observed movements, but from the other component of mobility: the ability to move.

An actor's ability to move does not necessarily correspond to the speed potentials to which he/she has access. To establish this state of affairs means to focus on the way in which actors construct their movements from technical networks, and therefore to focus on individual and collective trade-offs, constraints and appropriations of these technical systems. The notion of motility makes it possible to describe and analyse an actor's ability to move. Motility can be defined as the way in which an individual or group takes on the field of possible mobility and uses it to develop projects.

Motility may or may not be transformed into displacement, especially, it may be transformed into displacement in different ways, ways which mix the different forms of mobility. These forms are intertwined and each refer to specific social temporalities: day and week for daily mobility, month and year for travel, year and life cycle for residential mobility and life history for migration. These different forms have reciprocal impacts on each other. With this approach, we consider mobility as a single phenomenon that can manifest itself in different ways, which reconstitutes a unity where the puzzle of research fields and scientific disciplines had scattered the pieces.

The use of the notion of motility makes it possible to renew the way we look at mobility by highlighting strong and new logics. This is the case of the research carried out by Bruna Vendemmia in this book, which makes it possible to identify several original features of the great contemporary mobility:

### *Motility is used by many actors to avoid irreversibility and to push back spatial and temporal boundaries*

Bruna Vendemmia's work highlights that the juggling between modes of travel (car, plane, foot …) and between forms of travel (physical, virtual, telephone, SMS …) is used by people as a resource to repel the spatio-temporal incompatibilities faced by actors. This juggling, often quite inventive, implies extensive accessibility to transport and telecommunication networks, organisational skills and an appropriation of transport times and places. In dual-income couples, for example, practising long distance commuting by working in the train and some days at home connected to the Internet is becoming common. Different behavioural logics have thus been highlighted, each of which involves a degree of arbitration in which the actor seeks to optimise his mobility in relation to his various projects, in particular professional and family projects. Despite this, the travel solutions adopted are generally still considered by those concerned to be sub-optimal, but are accepted because they preserve an open field of possibilities. This apparent contradiction results from the fact that in trade-offs, mobility is secondary to the willingness to carry out activities or to establish a residential base. In fact, in many cases, actors avoid choosing between alternatives and seek to combine alternative terms through motility. Motility is therefore used to avoid irreversibility.

*The use of transport modes is not interchangeable, it is associated with specific motives.*

It is not enough to have implemented a high-speed train line for all those who live nearby and who would have an 'interest in using it' to use it. The results of Bruna Vendemmia's research show that the different uses of spaces consolidate in lifestyles through the spatial and temporal habits of daily life. Thus, the use of speed potential, such as that provided by a road network or a public transport network, produces specific forms of mobility, because this use mobilises skills and provides specific accessibility. If mobility produces uses in other areas of daily life, uses produce mobility in return.

*Motility is a resource that has the function of staying largely in a state of potential.*

Generally speaking, the actors seek to achieve the widest possible mobility by acquiring skills and access to the most numerous technical systems. Above all, they highlighted that from the point of view of the actors, motility does not necessarily have the objective of transforming itself into a movement.

Many actors acquire access and skills not to be mobile, but as insurance against risks of all kinds, as security to avoid being caught off guard in a wide variety of situations ranging from daily life to professional careers. The survey on the logic of action underlying modal practices, for example, showed that the appropriation of certain speed potentials is designed to remain in a state of latency: for many respondents, having access to a network broadens the spectrum of potential mobility, even if this access is not transformed into use. In particular, we have encountered this type of behaviour with regard to public transport; good access to their network is sought 'just in case'.

The notion of motility brings mobility into a new light: through the results presented, it responds to a dual logic of integration and disengagement in the sense of Hirschmann, which makes it a form of capital. While on the one hand mobility is becoming essential to reconcile increasingly numerous and spatially fragmented spheres of activity and projects, it also indicates a refusal by stakeholders to embark on irreversible paths. Indeed, many people play with the accessibility provided by technical systems to avoid having to choose, developing new organisational skills and appropriations of transport and communication systems, appropriations that divert these technical systems from the functions for which they were designed.

Regulating the participation/disengagement relationship highlighted by motility is a central issue for territorial development. Facilities, infrastructure and support for local and regional development that are intended to ensure harmonious and equitable development of the territory are currently confronted with behaviours that are not really integrated into spatial planning. Where territorial development plans seek to create balances, socially well-endowed actors produce imbalances by putting infrastructure, equipment and public support for local development at the service of their individual projects.

# Bibliography

Ahas, R. (2011). Mobile Positioning. In Büscher M., Urry, J. and Witchger, K. (eds), *Mobile Methods*, pp. 183–199. Abingdon: Routledge.

Amin, A. and Thrift, N. (2002). *Cities Reimagining the Urban*. Oxford: Polity Press and Blackwell Publisher Ltd.

Anderson, L. (2006). Analytic Autoethnography. *Journal of Contemporary Ethnography* 35(4), 373–395. doi:10.1177/0891241605280449.

Ascher, F. (2005a). La Métaphore Est un Transport. Des Idées Sur le Mouvement au Mouvement des Idées. *Cahiers Internationaux de Sociologie*, 118, 37–54. Paris: PUF. doi:10.3917/cis.118.0037.

Ascher, F. (2005b). *Les sens du movement: modernités et mobilités*. In Allemand, S., Ascher F. and Lévy, J. (eds), *Les Sens du Mouvement: Modernités et Mobilités*, pp. 21–34. Paris: Belin.

Bassand, M. and Brulhardt, M.C. (1980). *Mobilité Spatiale*. Saint-Saphorin: Georgi.

Bassand, M. and Brulhardt, M.C. (1983). La Mobilité Spatiale: un Processus Social Fondamental. *Espace, Populations, Sociétés, 1983–1981. Objectifs et champs d'étude*, 49–54. doi:10.3406/espos.1983.902.

Beria, P. and Grimaldi, R. (2011). An Early Evaluation of Italian High Speed Rail Projects. *Tema. Journal of Land Use Mobility and Environment* 4(3), 15–28. Available from: http://www.tema.unina.it/index.php/tema. doi:10.6092/1970-9870/486.

Beria, P. (2018). *Atlante dei Trasporti Italiani. Infrastrutture Offerta Domanda*. Novara: Libreria Geografica.

Bertolini, L. and Dijst, M. (2003). Mobility Environments and Network Cities. *Journal of Urban Design* 8(1), 27–43. doi:10.1080/1357480032000064755.

Bissel, D. (2009) Conceptualising Differently-Mobile Passengers: Geographies of Everyday Encumbrance in the Railway Station. *Social and Cultural Geography* 10, 173–195. doi:10.1080/14649360802652137.

Bissel, D. (2018). *Transit Life: How Commuting Is Transforming Our Cities*. In Urban and Industrial Environments series. Cambridge: The MIT Press.

Bissel, D., Vannini, P. and Jensen, O.B. (2017). Intensities of Mobility: Kinetic Energy, Commotion and Qualities of Supercommuting. *Mobilities* 2(6), 795–812. doi:10.1080/17450101.2016.1243935.

Boeri, S. (2011). *L'Anticittà*. Bari: Laterza.

Bonfiglioli, S. (2001). Urbanistica dei tempi: una nuova disciplina e un progetto formativo per l'Europa. *Territorio* 18, 30–32.

Bourdin, A. (2005). Les mobilités et le programme de la sociologie. *Cahiers Internationaux de Sociologie* 118(1), 5–21. doi:10.3917/cis.118.0005.
Bourdin, A. (2013). *La Ville Réversible: une Question d'Épistémologie Pour l'Urbanism.* In Scherrer, F. and Vanier, M. (eds), *Villes, Territoires, Réversibilités*, pp. 185–201. Paris: Hermann Editeur.
Büscher, M. and Urry, J. (2009). Mobile Methods and the Empirical. *European Journal of Social Theory* 12(1), 99–116. doi:10.1177/1368431008099642.
Büscher, M., Urry, J. and Witchger, K. (2011). *Mobile Methods.* Abingdon: Routledge.
Castells, M. (1996). *The Rise of the Network Society, the Information Age: Economy, Society and Culture.* Vol. I. Cambridge, MA; Oxford, UK: Blackwell.
Colleoni, M. and Zajczyc, F. (2003). La Mobilità nello Spazio e nel Tempo delle Popolazioni Urbane. Una Lettura Sociologica. *Sociologia Urbana e Rurale* 72, 35–54.
Comune di Milano. (2014). *Gli esiti della giornata del lavoro agile. Una sperimentazione del Piano Territoriale degli orari di Milano [the results of Milan smart working day: an experiment of Milan time working plan]* Available from: https://smartforexpo.files.wordpress.com/2015/04/esiti-della-giornata-del-lavoro-agile-2014-2.pdf [accessed 5 December 2019]
Cresswell, T. (2006). *On the Move: Mobility in the Modern World.* Abingdon: Routledge.
Cresswell, T. (2010a). Towards a Politics of Mobility. *Environmental and Planning D. Society and Space* 28(1), 17–31. doi:10.1068/d11407.
Cresswell, T. (2010b). Mobilities I: Catching Up. *Progress in Human Geography* 35(4), 550–558. doi:10.1177/0309132510383348.
Crosta, P.L. (2003). Reti Translocali. Le Pratiche d'Uso del Territorio Come 'Politiche' e come 'Politica'. *Foedus* 7, 5–18.
Crosta, P.L. (2010). *Pratiche. Il Territorio è l'Uso che Se ne Fa.* Milano: Franco Angeli.
Deleuze, G. and Guattari, F. (2010). *Nomadology: The War Machine.* Seattle: Wormwood Distribution. Originally appearing in *A Thousand Plateaus* (1980) Paris: Minuit.
Di Marino, M. and Lapintie, K. (2017). Emerging Workplaces in Post-Functionalist Cities. *Journal of Urban Technology* 24(3), 5–25. doi:10.1080/10630732.2017.1297520.
Duchêne-Lacroix, C. (2013). Caractérisation des Situations d'Immobilité: Réflexions Methodologiques Elements Pour une Typologie des Pratiques Plurirsidentielles et d'un Habiter Multilocal. *E-Migrinter - Et l'Immobilité dans la circulation?* Laboratoire MIGRINTER, 11, 151–167. Available from: https://journals.openedition.org/e-migrinter/323 [2 December 2019] doi:10.4000/e-migrinter.323.
Elliott, A. and Urry, J. (2010). *Mobile Lives.* Abingdon: Routledge.
European Charter for the Development of Social and Societal Initiatives in Train Stations (2010) Available from: http://www.garesolidaire.net/attachments/Chartesig2__EN__4_.pdf [accessed 2 December 2019].
Ferrio, S. (2013). *Da Zero a Italo. Così è nata la concorrenza.* Milano: Skyra.
Frug, G.E. (1999). *City Making: Building Communities without Building Walls.* Princeton, NJ: Princeton University Press.
Geels, F.W. (2012). A Socio-Technical Analysis of Low-Carbon Transitions: Introducing the Multi-level Perspective into Transport Studies. *Journal of Transport Geography* 24, 471–482. doi:10.1016/j.jtrangeo.2012.01.021.
Gwiazdzinski, L. (2013). De l'Hypothèse de Reversibilité, à la Ville Malléable et Augmentée: Vers un Neo-Situtionnisme. In Scherrer, F. and Vanier, M. (eds), *Villes, Territoires, Réversibilités*, pp. 205–219. Paris: Hermann Editeur.

Haesbaert da Costa, R. (2005). De la Déterritorialisation à la Multiterritorialité. In Allemand, S., Ascher, F. and Lévy, J. (eds), *Les Sens du Mouvement: Modernités et Mobilités*, pp. 69–79. Paris: Belin.

Hågerstrand, T. (1970). What about People in Regional Science?. *Papers of the Regional Science Association* 24(1), 6–21. doi:10.1007/BF01936872.

IRES, Istituto per le Ricerche Economiche e Sociali dell'UGL. (2013). Gli effetti della crisi sul lavoro in Italia maggio 2013. Available from: https://www.fondazionedi-vittorio.it/it/effetti-della-crisi-sul-lavoro-italia-%E2%80%93-maggio-2013 [accessed 04 December 2019].

ISFORT. Istituto Superiore di Formazione e Ricerca sui Trasporti (2018). *15 °rapporto sulla mobilità degli Italiani.* Available from: https://www.isfort.it/wp-content/uploads/2019/09/Rapporto_Mobilita_2018.pdf [accessed 04 December 2019].

ISTAT (2013a) *Verde Urbano [Urban green].* Available from: https://www.istat.it/it/files//2013/04/Verde-urbano.pdf [accessed 04 December 2019].

ISTAT (2013b) La superficie dei comuni, delle provincie e delle regioni Italiane [the surface of municipalities, Provinces and Region in Italy]. *15 °Censimento Cenerale della Popolazione e delle Abitazioni.* Available from: https://www.istat.it/it/files/2015/04/Superfici-delle-unit%C3%A0-amministrative-Testo-integrale.pdf [accessed 04 December 2019].

ISTAT (2018). Natalità e fecondità della popolazione residente [Nativity and fecondity of resident population]. Available from: https://www.istat.it/it/files//2019/11/Report_natalit%C3%A0_anno2018_def.pdf [accessed 04 December 2019].

Jain, J. and Lyons, G. (2008). The gift of travel time. *Journal of Transport Geography* 16, 81–89. doi:10.1016/j.jtrangeo.2007.05.001.

Kayden, J. S. (2000). *Privately Owned Public Space: The New York City Experience.* Hoboken, NJ: John Wiley and Sons. .

Kaufmann, V. (2002). *Re-thinking Mobility.* Aldershot: Ashgate.

Kaufmann, V. (2005). Mobilités et Réversibilités: Vers des Sociétés plus Fluides? *Cahiers Internationaux de Sociologie*, 1(118), 119–135. Paris: PUF. doi:10.3917/cis.118.0119.

Kaufmann, V. (2011). *Re-thinking The City.* Lausanne: EPFL Press.

Kaufmann, V. (2012). Déplacement. *Forum Vies Mobiles.* Available from: http://fr.forumviesmobiles.org/reperes/deplacement-452.

Kaufmann, V. (2016). Putting Territory to the Test of Reversibility. In Pucci, P. and Colleoni, M. (eds), *Understanding Mobilities for Designing Contemporary Cities* (pp. 35–48). Heidelberg-New York-Dordrecht-London: Springer.

Kesselring, S. (2006). Pioneering Mobilities: New Patterns of Movement and Motility in a Mobile World. *Environment and Planning A* 38(2), 269–279. doi:10.1068/a37279.

Kesselring, S. and Vogl, G. (2008). Networks, Scapes and Flows – Mobility Pioneers between First and Second Modernity. In Canzler, W., Kaufmann V. and Kesselring, S. (eds), *Tracing Mobilities. Toward a Cosmopolitan Perspective* pp. 163–179. Aldershot: Ashgate.

Lanzani, A. and Granata, E. (2011). Metamorfosi dell'Abitare. In Lanzani, A. (ed.), *In Cammino nel Paesaggio. Questioni di Geografia e di Urbanistica*, pp. 183–202. Milan: Carocci.

Larsen, J., Axhausen, K. W. and Urry, J. (2006). *Geographies of Social Networks: Meetings, Travel and Communications, in Mobilities* 1(2), 261–283. doi:10.1080/17450100600726654.

Latham, A. (2003). Research, Performance, and Doing Human Geography: Some Reflections on the Diary-Photograph, Diary-Interview Method. *Environment and Planning A* 35, 1993–2017. doi:10.1068/a3587.

Latour, B. (2005). *Reassembling the Social: An Introduction to Actor-Network Theory.* Oxford: Oxford University Press.

Laurier, E. (2004). Doing Office on the Motorway. *Theory Culture Society* 21, 261–277. doi:10.1177/0263276404046070.

Laurier, E. (2010). Participant Observation. In Clifford, N., French, S. and Valentine, G. (eds), *Key Methods in Geography*2nd edn, pp. 116–130. London: Sage Publications.

Law, J. and Urry, J. (2004) Enacting the Social. *Economy and Society*, 33:3, 390–410 doi:10.1080/0308514042000225716.

Lefebvre, H. (2004). *Rhythmanalysis: Space, Time and Everyday Life.* London, New York: Continuum.

Lévy, J. (1994). *L'espace Légitime.* Paris: Presses de la fondation nationale des sciences politiques.

Lucas, K. (2012). Transport and Social Exclusion: Where Are We Now? *Transport Policy* 20, 105–113. doi:10.1016/j.tranpol.2012.01.013.

Lyons, G. and Urry J. (2005). Travel Time Use in the Information Age. *Transportation Research Part A: Policy and Practice* 39, 257–276. doi:10.1016/j.tra.2004.09.004.

Lyster, C. (2013). Infrastructural Cartography: Drawing the Space of Flows. In Sen, A. and Johung, J. (eds), *Landscapes of Mobility. Culture, Politics, and Placemaking*, pp. 239–253. Aldershot: Ashgate.

Martens, K. (2017). *Transport Justice: Designing Fair Transportation Systems.* Abingdon: Routledge.

Massey, D. (1994). *Space, Place and Gender.* Minneapolis: University of Minnesota Press.

Melucci, A. (1991). *Il gioco dell'Io.* Milano: Feltrinelli.

Mitchell, D. (2003). *The Right to the City. Social Justice and the Fight for Public Spaces.* New York, London: The Guilford Press.

Moscarini, L. V. (2009) *Diritto Privato e Interessi Pubblici, Saggi di Diritto Civile 2001–2008*. Milano: Wolter Kluwer Italia s.r.l.

Nadler, R. (2014). *Plug&Play Places. Lifeworlds of Multilocal Creative Knowledge Workers.* Munich: De Gruyter Open. Available from: https://www.degruyter.com/view/product/447763?format=EBOK.

OECD (2014) Society at a Glance 2014 Highlights: Italy the Crisis and Its Aftermath. Available from: https://www.oecd.org/els/societyataglance.htm [accessed 2 December 2019].

Oldenburg, R. (1989). *The Great Good Place: Cafes, Coffee Shops, Community Centers, Beauty Parlors, General Stores, Bars, Hangouts, and How They Get You Through the Day.* New York: Paragon House.

Öhman, M. and Lindgrend, U. (2003). Who Is the Long-distance Commuter? Patterns and Driving Forces in Sweden. *Cybergeo: European Journal of Geography*, 243. doi:10.4000/cybergeo.4118.

Pasqui, G. (2008). *Città, Popolazioni, Politiche.* Milano: editoriale Jaca book SpA.

Pradel, B. (2015). *Du long terme de l'ancrage au court terme de la mobilité: systéme de lieux-moments et échelles temporelles de l'habiter.* 4émes Rencontres Scientifiques Internationales de la Cité des Territoires, Grenoble, 25 -27 Marsch.

Pucci, P. (1996). *I Nodi Infratsrutturali: Luoghi e non Luoghi Metropolitani.* Milano: Franco Angeli,

Pucci, P. (2016). Mobility practices as a knowledge and design tool for urban policy. In Pucci, P. and Colleoni, M. (eds), *Understanding Mobilities for Designing Contemporary Cities.* Berlin: Springer. doi:10.1007/978-3-319-22578-4_1.

Pucci, P. (2019). Emerging Mobilities: New Practices, New Needs. In Pucci, P. and Vecchio, G. (eds), *Enabling Mobilities. Planning Tools for People and Their Mobilities.* Cham: Springer. doi:10.1007/978-3-030-19581-6.

Pucci, P. and Manfredini, F. (2011). *Seguire le Tracce. Se e Come i Dati di Telefonia Mobile Consentono di Descrivere l'Uso della Città.* XIV SIU Conference, Conference Proceedings, Abitare l'Italia Territori, Economie, disuguaglianze. Turin, March.

Pucci, P. and Vecchio, G. (2019). *Enabling Mobilities. Planning Tools for People and Their Mobilities.* Cham: SpringerBriefs. doi:10.1007/978-3-030-19581-6.

Pucci, P. and Vecchio, G. (2019). Trespassing for Mobilities. Operational Directions for Addressing Mobile Lives. *Journal of Transport Geography* 81, 1–11 doi:10.1016/j.jtrangeo.2019.102536.

Proust, M. (1913). *Du côté de chez Swann.* Paris: Grasset.

Raffestin, C. (1986). Territorialité: Concept ou Paradigme de la géographie sociale? *Geographica Helvetica* 2, 91–96. doi:10.5194/gh-41-91-1986.

Ralph, D. (2014). *Work, Family and Commuting in Europe: The Lives of Euro-Commuters.* London: Palgrave.

Ravalet, E., Vincent-Geslin, S. and Dubois, Y. (2014) *High Mobility…to have a job?* XIV STRC 2014, Swiss Transport Research Conference. Monte Verità Ascona, May 2014. Available from: http://www.strc.ch/conferences/2014/Ravalet_EtAl_2.pdf [03 December 2019]

Sandow, S. and Westin, K. (2010). The Persevering Commuter – Duration of Long-distance Commuting. *Transportation Research Part A: Policy and Practice* 44, 433–445. doi:10.1016/j.tra.2010.03.017.

Schneider, N. F. and Meil, G. (eds). (2008). *Mobile Living Across Europe I: Relevance and Diversity of Job-related Spatial Mobility in Six European Countries.* Opladen: Barbara Budrich Publishers.

Sen, A. and Johung J. (2013). *Landscapes of Mobility. Culture, Politics, and Place-making.* Aldershot: Ashgate.

Schönfelder, S. and Axhausen, K. W. (2003). Activity spaces: measures of social exclusion? *Transport Policy* 10, 273–286. doi:10.1016/j.tranpol.2003.07.002.

Sevtsuk, A. and Ratti, C. (2010). Does Urban Mobility Have a Daily Routine? Learning from the Aggregate Data of Mobile Networks. *Journal of Urban Technology* 17(1), 41–60. doi:10.1080/10630731003597322.

Sheller, M. (2011). Mobility. *Socepodia.isa.* Available from: http://www.sagepub.net/isa/resources/pdf/Mobility.pdf [02 December 2019]. doi:10.1177/205684601163.

Sheller, M. and Urry, J. (2006) The New Mobilities Paradigm. *Environment and Planning A*, 38, 207–226. doi:10.1068/a37268.

Société du Grand Paris (2012) Architecture and Design des Gares du Grand Paris. Résultats de la consultation de la mission de conseil. Available from:http://www.societedugrandparis.fr/wp-content/uploads/2014/05/dp-26-06-2012.pdf [accessed 06 December 2019].

Soja, E. (2010). *Seeking Spatial Justice.* Minneapolis: University of Minnesota Press.

Stock, M. (2006) L'hypothèse de l'habiter poly-topique: pratiquer les lieux géographiques dans les sociétés à individus mobiles. *EspacesTemps.net.* Available from: https://www.espacestemps.net/articles/hypothese-habiter-polytopique/ [2 December 2019]

Stock, M. (2013) Politopie. In Lévy, J. and Lussault, M. (eds), *Dictionnaire de la Géographie et de l'Espace des Sociétés.* Paris: Belin, pp. 794–796.

Tarrius, A. (2000). *Les nouveaux cosmopolitisme. Mobilités, Identités, Territoirés.* La Tour d'Aigues: Editions de l'aube.

Urry, J. (2000). *Sociology Beyond Societies.* New York: Routledge.
Urry, J. (2007). *Mobilities.* Cambridge, UK: Polity.
Vannini, P. (2009) *The Cultures of Alternatives Mobilities. Routes Less Travelled.* Aldershot: Ashgate.
Van Schaick, J. and Van der Spek, S.C. (2008). *Urbanism on Track. Application of Tracking Technologies in Urbanism.* Delft: Delft University Press.
Vendemmia, B. (2018). Reversible mobilities and territories to improve urban synergies. In D'arienzo, R. and Younès, C. (eds) *Synergies urbaines. Pour un métabolisme collectif des villes*, pp. 263–272. Genève: Metispresses.
Viesti, G. (2019). Il trasporto ferroviario in Italia nel XXI secolo: un paese sempre più diseguale. *Economia e Politica* 11(8), 2. Available from: https://www.economiaepolitica.it/2019-anno-11-n-18-sem-2/il-trasporto-ferroviario-in-italia-nel-xxi-secolo-un-paese-sempre-piu-diseguale/ [03 December 2019].
Viganò, P. (2010). *Territorio dell'Urbanistica. Il Progetto come Produttore di Conoscenza.* Roma: Officina.
Viganò, P. (2013). The Horizontal Metropolis and Gloeden's Diagrams Two Parallel Stories. *Oase* 89, 94–11. OASE Foundation & NAi Publishers. Available from: https://www.oasejournal.nl/en/Issues/89/TheHorizontalMetropolisAndGloedensDiagrams [03 December 2019].
Vincent-Geslin, S. and Ortar, N. (2012). De la Mobilitè au Racines. In Vincent-Geslin, S. and Kaufmann, V. (eds), *Mobilité sans Racine. Plus Loin, Plus Vite…Plus Mobiles?*, pp. 51–64. Paris: Descartes and Cie.
Viry, G. and Kaufmann, V. (eds) (2015). *High Mobility in Europe. Work and Personal Life.* Basingstoke: Palgrave Macmillan.
Viry, G., Ravalet, E. and Kaufmann, V. (2015). High Mobility in Europe: An Overview. In Viry, G. and Kaufmann, V. (eds), *High Mobility in Europe. Work and Personal Life.* Basingstoke: Palgrave Macmillan.
Widmer, E. D. (2006). Who Are My Family Members? Bridging and Binding Social Capital in Family Configurations . *Journal of Personal and Social Relationships* 23 (6), 979–998. doi:10.1177/0265407506070482.
Zajczyk, F. (2000). *Tempi di Vita e Orari della Città. La Ricerca Sociale e il Governo Urbano.* Milano: Franco Angeli.

# Index

For Product Safety Concerns and Information please contact our EU
representative GPSR@taylorandfrancis.com
Taylor & Francis Verlag GmbH, Kaufingerstraße 24, 80331 München, Germany

www.ingramcontent.com/pod-product-compliance
Ingram Content Group UK Ltd.
Pitfield, Milton Keynes, MK11 3LW, UK
UKHW022025061126
477531UK00002B/38
*9780367406400*